Strategic Management of Sustainable Urban Development

Economic Downturns, Metropolitan Governance and Local Communities

RIVER PUBLISHERS SERIES IN CHEMICAL, ENVIRONMENTAL, AND ENERGY ENGINEERING

Indexing: All books published in this series are submitted to the Web of Science Book Citation Index (BkCI), to SCOPUS, to CrossRef and to Google Scholar for evaluation and indexing.

The "River Publishers Series in Chemical, Environmental, and Energy Engineering" is a series of comprehensive academic and professional books which focus on Environmental and Energy Engineering subjects. The series focuses on topics ranging from theory to policy and technology to applications.

Books published in the series include research monographs, edited volumes, handbooks and textbooks. The books provide professionals, researchers, educators, and advanced students in the field with an invaluable insight into the latest research and developments.

Topics covered in the series include, but are by no means restricted to the following:

- Energy and Energy Policy
- Chemical Engineering
- Water Management
- Sustainable Development
- Climate Change Mitigation
- Environmental Engineering
- Environmental System Monitoring and Analysis
- Sustainability: Greening the World Economy

For a list of other books in this series, visit www.riverpublishers.com

Strategic Management of Sustainable Urban Development

Economic Downturns, Metropolitan Governance and Local Communities

Sabato Vinci

University of Roma Tre, Italy

Luca Salvati

University of Macerata, Italy

River Publishers

Published, sold and distributed by:
River Publishers
Alsbjergvej 10
9260 Gistrup
Denmark

www.riverpublishers.com

ISBN: 978-87-7022-166-5 (Hardback)
978-87-7022-165-8 (Ebook)

Contents

Preface

Management Studies and Public Governance Perspectives for Effective and Sustainable Urban Development

The present work introduces a novel interpretation of the intimate relationship between urban studies and management and public governance studies, for strategic development and management of metropolitan and environmental systems. The linkage between management and other areas of scientific knowledge finds fertile ground in the Italian school of Business Economics, founded by Gino Zappa. This "father" of matter taught us the importance of factual analysis as a baseline for each business-economic reflection and also took ideas from the American institutional economics (Veblen, North, and above all, Commons), coming to suggest a "broad" approach to the study of business issues. So, business and management sciences can progress in the research about efficient governance theories and management models, taking advantage of contribution from other sciences for an accurate factual analysis, to achieve integrated and effective solutions.

So, the integration between business studies and other scientific fields is inherent to the nature of management, which is in itself a field of knowledge always "inclusive" of different sensitivities and contents, with a focus on results. This attitude of management is essential above all to apply the principles of economics and business sciences to public organisations, where stakeholders have traditionally a stronger role than in the private sector. Furthermore, this is a great advantage for those areas of social organisation where it is important to include technical and extra-economic knowledge of business to develop efficient governance theories and management models. This is typically the case of urban management and local development, but also, for example, of management of healthcare organisations, as well as of emergency and natural disasters management. In particular, the complexity of the issue of local and urban development requires a greater capacity by policy makers and public administrators to interpret deep relationships that exist between business models, socioeconomic traditions, geographical and landscape characteristics of territory, organisational structures and management skills.

As is known, the classical economic debate has traditionally focused mainly on the issue of economic growth, putting in second place the reflection about limits connected with territorial exploitation. There are limits also in the business and management sciences, although to a lesser extent: in fact, in the Italian tradition of Business Economics, the theoretical elaboration of the "organisations" (in italian "aziende") was based on the concept of "direct" or "indirect" satisfaction of human needs, giving rise to the traditional theoretical models of "supply organisations" and "production organisations" developed by Gino Zappa and his academic school. Therefore, if from a business economics point of view we consider generally the organisation as "economic coordination for satisfaction of human needs", it follows that the organisations, and in particular public organisations, must be able to interpret both old then new stakeholders' needs, by offering a governance response connected with a proper socioeconomic development strategy. Today, public institutions are showing increasing attention to sustainability issues, and citizens are increasingly sensitive to the idea of sustainable socioeconomic development. Therefore, the great challenge of modern public governance is to identify models of technology and organisation suitable to meet the current human needs without compromising the ability of future generations to find equal (and possibly better) satisfaction of their own needs. The academic debate took into account this evolution of the relationship between policies, economy, and sustainability, and in many aspects, it preceded and encouraged this evolution. Thus, the question of management not only efficient but also sustainable of development process – with particular reference to the issues of urban transformation, climate changes, and exploitation of natural resources – becomes of primary importance in the debate in modern management studies. Further, this trend also involves private enterprises that are nowadays recipients of increasing institutional incentives to maintain ethical and social responsibility behaviors.

This book captures these topics in their entirety, and it appreciably declines them in relation to concrete reality, by showing great scientific and methodological rigor. Thus, the result of this notable academic work takes its place within a fundamental line in the modern elaboration of managerial sciences, providing a thorough analysis of economic downturn effects on metropolitan governance and local communities, as well as by suggesting some attractive management solutions for the implementation of effective and sustainable strategies for local and urban devel

Eugenio D'Amico

Full Professor of Business Economics and Director of the Research Laboratory of Economics, Governance and Ethics of Organisations (LEGEA), University of Roma Tre, Italy

List of Figures

List of Tables

1

Beyond the Crisis: From Smart Cities to Smart Landscapes

Luca Salvati

This book illustrates different aspects of local development and their governance models in complex socioeconomic and environmental systems (Salvati et al., 2017; Serra et al., 2018). Assuming recession as the starting point for policy challenge, economic downturns were interpreted as opportunities for change reshaping society, landscapes, and the latent mechanisms of regional growth (Florida, 2011; Zambon et al., 2018, 2019; Salvati et al., 2019). Coming from diverse academic backgrounds, our book addresses paradigmatic visions about regional and urban dynamics, focusing on landscape transformations and socioeconomic disparities (Prokopovà? et al., 2018). A specific approach based on a direct investigation of mechanisms of local development, cultural, and environmental values within a strategic territorial vision (Chelli et al., 2016; Gigliarano and Chelli, 2016), was finally proposed. More specifically, the idea that a "smart city" is not only economically attractive or competitive, but also socially cohesive and environmentally sustainable, was introduced and commented extensively in this book. A "smart city" means a city where cooperation matters more than competition; a city where identity and diversity of places are more important than the mere imitation of models conceived and tested somewhere else; a city where the proper value of landscape – in terms of cultural and environmental goods – is the real challenge for communities going toward a truly "smart" transition.

Resilient districts become the dimension where management, planning, and projects are more adapted to the needs of local communities, considering social diversity, landscape peculiarity, and local identity as specific factors of resistance to crisis (Carlucci et al., 2018; Cecchini et al., 2018; Morrow et al., 2018). Moreover, resilient districts represent a socioeconomic model alternative to standardized growth paths based on the intrinsic competitiveness between economic, political, and social actors. Local competitiveness led to spatial policies and best practices often failing to address the real attitudes of local communities, imposing decontextualized objects at the same time (Bristow, 2009; Purcell, 2009; Morelli et al., 2014). Decontextualized types of development might turn to be problematic in many ways, from the strictly economic point of view – creating working conditions depending on external contexts and forces – to the ecological point of view, imposing growth targets that do not incorporate environmental issues (Cuadrado Ciuraneta et al., 2017).

With a context based in a "triple crunch" (austerity policies, climate changes, and increases in oil prices), to consolidate an alternative to "competitive models" of local development is imperative (Carlucci et al., 2017). Resilience theory provides the appropriated knowledge and informs local development built on local-based concepts, instead of focusing on competitiveness factors only (Biasi et al., 2019). The purpose of these policies is to "*create more robust economic and social spaces by empowering producers and consumers to interact locally, seeking to reduce dependency upon distant and larger scale agents, namely nonlocal and large corporations and the nation state*" (Leitner et al., 2007).

Going back to the local scale, abandoning interpretative paradigms oriented toward the logic of "global networks" is intrinsically linked with the idea of socioeconomic resilience. While resilience appears as a key issue in socioeconomic thought, strategic planning and landscape design (Cecchini et al., 2019), it has been properly considered by institutions and policy makers only in some cases (e.g. Chelli et al., 2009). Dealing with resilience requires a deep analysis of societies, institutions, and local contexts, understanding the resilient nature beyond the system. This "holistic" concept makes research on local factors of resilience even more difficult (Adger, 2000).

Resilient systems rely upon peculiarities and resources to restart in case of sudden changes. In other words, resilience could be defined as the "*region's ability to experience positive economic success that is socially inclusive, works within environmental limits and which can ride global economic purchase*" (Ashby et al., 2009). Resilience of local districts should be therefore understood as the ability to adapt to economic, technological, and political changes, which affect evolutionary dynamics and trajectories pursued by regional economies (Simmie and Martin, 2009).

The main argument of this book allows to think about landscape management as a resilient strategy for local districts (Masini et al., 2019; Bertini et al., 2019; Marchi et al., 2018; Fabbio et al., 2018), promoting diversification of economic activities and other distinctive features, that are less tied to production and more grounded on social, cultural, and identity qualities (Di Feliciantonio et al., 2018; Carlucci et al., 2019). Truly resilient assets can be obtained by putting together the economic characteristics of a given area with its own cultural attitudes. Landscape approaches envision the perspective of sustainable development within territories by recognizing identity values and local culture, promoting actions enhancing the specific features of places and engaging stakeholders into dedicated conservation strategies (Pili et al., 2017). Territorial processes stimulate social cohesion and community identity, promoting tourism and regional marketing (Carlucci et al., 2018; Ciommi et al., 2017, 2019). A landscape-oriented planning strategy leads to recognize the identity of places and the values preserved by local communities. In this way, local districts could find a new life even after a period of crisis, in order to pursue the development of a truly sustainable society (Carlucci et al., 2018; Zambon et al., 2018; Cecchini et al., 2019; Salvati et al., 2019).

In a postcrisis period, landscape might frame agriculture, economic activities, and the existing socio-demographic relationships in a context where development issues relate to sustainable approaches and local backgrounds (Salvati, 2018; Ciommi et al., 2018; Zambon et al., 2017; Serra et al., 2018). In these regards, peri-urban areas might turn into "multifunctional spaces," favoring traditional economies and land uses related to agriculture. The main spatial assumption of sustainable development deals with the equilibrium between human pressure and the use of land resources, seeking to ensure a balance between human needs and the available natural capital (Salvati et al., 2019). Sustainable development underlines the need of protecting land resources when a regional transformation occurs, in order to safeguard and improve the quality of future environments, by integrating production and consumption dynamics with the value of solidarity.

References

Adger, W. N. (2000). Social and ecological resilience: Are they related? *Progress in Human Geography*, 24, 347.

Ashby, J., Cox, D., McInroy, N. (2009). An International Perspective of Local Government as Steward of Local Economic Resilience. Report by the Centre for Local Economic Strategies: Manchester.

Bertini, G., Becagli, C., Chiavetta, U., Ferretti, F., Fabbio, G., Salvati, L. (2019). Exploratory analysis of structural diversity indicators at stand level in three Italian beech sites and implications for sustainable forest management. *Journal of Forestry Research*, 301, 121–127.

Biasi, R., Brunori, E., Ferrara, C., Salvati, L. (2019). Assessing impacts of climate change on phenology and quality traits of *Vitis vinifera* L.: The contribution of local knowledge. *Plants*, 85, 121.

Bristow, G., Wells, P. (2005). Innovative discourse for sustainable local development: A critical analysis of eco-industrialism., *International Journal of Innovation and Sustainable Development*, 1, 168–179.

Brown, J., Dillard, J. (2013). Agonizing over engagement: SEA and the "death of environmentalism" debates. *Critical Perspectives on Accounting*, 24(1), 1–18.

Carlucci, M., Chelli, F.M., Salvati, L. (2018). Toward a new cycle: Short-term population dynamics, gentrification, and re-urbanization of Milan (Italy). *Sustainability (Switzerland)*, 10(9), 3014.

Carlucci, M., Grigoriadis, E., Venanzoni, G., Salvati, L. (2019) Crisis-driven changes in construction patterns: Evidence from building permits in a Mediterranean city. *Housing Studies*, 338, 1151–1174.

Carlucci, M., Zambon, I., Salvati, L. (2019). Diversification in urban functions as a measure of metropolitan complexity. *Environment and Planning B: Urban Analytics and City Science*, 239980831982837.

Cecchini, M., Cividino, S., Turco, R., Salvati, L. (2019). Population age structure, complex socio-demographic systems and resilience potential: A spatio-temporal, evenness-based approach. *Sustainability* (Switzerland), 117, 2050.

Cecchini, M., Zambon, I., Pontrandolfi, A., Turco, R., Colantoni, A., Mavrakis, A., Salvati, L. (2019). Urban sprawl and the 'olive' landscape: Sustainable land management for 'crisis' cities. *GeoJournal*, 84(1), 237-255.

Chelli, F., Gigliarano, C., Mattioli, E. (2009). The impact of inflation on heterogeneous groups of households: An application to Italy. *Economics Bulletin*, 29(2), 1276-1295.

Chelli, F.M., Ciommi, M., Emili, A., Gigliarano, C., Taralli, S. (2016). Assessing the Equitable and Sustainable Well-Being of the Italian Provinces. *International Journal of Uncertainty, Fuzziness and Knowlege-Based Systems,* 24, 39-62.

Ciaccia, C., La Torre, A., Ferlito, F., Testani, E., Battaglia, V., Salvati, L., Roccuzzo, G. (2019). Agroecological practices and agrobiodiversity: A case study on organic orange in southern Italy. *Agronomy*, 92, 85.

Ciommi, M., Gigliarano, C., Emili, A., Taralli, S., Chelli, F.M. (2017). A new class of composite indicators for measuring well-being at the local level: An application to the Equitable and Sustainable Well-being (BES) of the Italian Provinces. *Ecological Indicators,* 76, 281-296.

Ciommi, M., Chelli, F. M., Carlucci, M., Salvati, L. (2018). Urban growth and demographic dynamics in southern Europe: Toward a new statistical approach to regional science. *Sustainability* (Switzerland), 108(1), 2765.

Ciommi, M., Chelli, F.M., Salvati, L. (2019). Integrating parametric and non-parametric multivariate analysis of urban growth and commuting patterns in a European metropolitan area. *Quality and Quantity*, 53(2), 957-979.

Di Feliciantonio, C., Salvati, L., Sarantakou, E., Rontos, K. (2018). Class diversification, economic growth and urban sprawl: Evidences from a pre-crisis European city. *Quality and Quantity*, 52(4), 1501–1522.

Fabbio, G., Cantiani, P., Ferretti, F., di Di Salvatore, U., Bertini, G., Becagli, C., Chiavetta, U., Marchi, M., Salvati, L. (2018). Sustainable land management, adaptive silviculture, and new forest challenges: Evidence from a latitudinal gradient in Italy. *Sustainability* (Switzerland), 107(1), 2520.

Florida, R. (2011). *The Great Reset: How New Ways of Living and Working Drive Postcrash Prosperity*. Florida: Random House Canada.

Gigliarano, C., Chelli, F.M. (2016). Measuring inter-temporal intragenerational mobility: an application to the Italian labour market. *Quality and Quantity*, 50(1), 89-102.

Leitner, H., Peck, J., Sheppard, E. S. (Eds.). (2007). *Contesting Neoliberalism: Urban Frontiers*. Guilford Press.

Marchi, M., Ferrara, C., Biasi, R., Salvia, R., Salvati, L. (2018). Agro-forest management and soil degradation in Mediterranean environments: Towards a strategy for sustainable land use in vineyard and olive cropland. *Sustainability* (Switzerland), 107(2), 2565.

Masini, E., Tomao, A., Barbati, A., Corona, P., Serra, P., Salvati, L. (2019).
Urban growth, land-use efficiency and local socioeconomic context: A comparative analysis of 417 metropolitan regions in Europe. *Environmental Management*, 633, 322–337.

Morrow, N., Salvati, L., Colantoni, A., Mock, N. (2018). Rooting the future: On-farm trees' contribution to household energy security and asset creation as a resilient development pathway-evidence from a 20-year panel in rural Ethiopia. *Sustainability* (Switzerland), 1012, 4716.

Prokopová, M., Cudlín, O., Vèeláková, R., Lengyel, S., Salvati, L., Cudlín, P. (2018). Latent drivers of landscape transformation in Eastern Europe: Past, present and future. *Sustainability* (Switzerland), 108, 2918.

Purcell, M. (2009). Resisting neoliberalisation: Communicative planning or counterhegemonic movements? *Planning Theory*, 8, 140–165.

Salvati, L. (2018). The 'niche' city: A multifactor spatial approach to identify local-scale dimensions of urban complexity. *Ecological Indicators*, 94(3), 62–73.

Salvati, L., De Zuliani, E., Sabbi, A., Cancellieri, L., Tufano, M., Caneva, G., Savo, V. (2017). Land-cover changes and sustainable development in a rural cultural landscape of central Italy: Classical trends and counterintuitive results. *International Journal of Sustainable Development & World Ecology*, 24(1), 27–36.

Salvati, L., Serra, P., Bencardino, M., Carlucci, M. (2019). Re-urbanizing the European city: A multivariate analysis of population dynamics during expansion and recession times. *European Journal of Population*, 35(1), 1–28.

Salvati, L., Ciommi, M.T., Serra, P., Chelli, F.M. (2019). Exploring the spatial structure of housing prices under economic expansion and stagnation: The role of socio-demographic factors in metropolitan Rome, Italy. *Land Use Policy*, 81, 143-152.

Serra, P., Saurí, D., Salvati, L. (2018). Peri-urban agriculture in Barcelona: outlining landscape dynamics vis-à-vis socio-environmental functions. *Landscape Research*, 43(5), 613–631.

Simmie, J., Martin, R. (2010). The economic resilience of regions: Towards an evolutionary approach. *Cambridge Journal of Regions, Economy and Society*, 3, 27–43.

Zambon, I., Benedetti, A., Ferrara, C., Salvati, L. (2018). Soil matters? A multivariate analysis of socioeconomic constraints to urban expansion in Mediterranean Europe. *Ecological Economics*, 146, 173–183.

Zambon, I., Colantoni, A., Carlucci, M., Morrow, N., Sateriano, A., Salvati, L. (2017). Land quality, sustainable development and environmental degradation in agricultural districts: A computational approach based on entropy indexes. *Environmental Impact Assessment Review*, 64, 37–46.

2

Management of Local Development and Socio-Economic Disparities: Crisis Versus Postcrisis Perspectives

Sabato Vinci and Luca Salvati

Two terms have been used for defining cities in the Latin world: "urbs" and "civitas". The term "urbs" is undoubtedly connected with the notion of "civilization", and the two words have the same base. However, the former notion specifically designates the urban structure of human societies (spatial patterns of urbanization and housing typologies) while the latter term refers to the humans that are part of the society. Therefore the "city" evokes both the notion of social life and the modalities with which humans inhabit the territory. In this sense, urbanization is considered as the expression of cultural and intellectual characteristics of a society. Urban development patterns evolve according to multiple variables. Above all, they are a product of social action, a human creation that varies enormously over time and space. The geographical characteristics of the area, the intrinsic culture of the inhabitants, market and social dynamics, the governing body of the nation, and relevant human/natural events all contribute to the formation of cities' structure. It is therefore unrealistic to speak about a unique urban development type as these elements vary significantly between countries, e.g. of the same continent (Martinotti, 1993). However, it is correct to argue that, each type of urban development, sooner or later poses the need for management of the socioeconomic development of the territory (Boje et al., 2000; Savall, 2003). With reference to the time when the government's planning action is carried out, it may depend on historical factors of individual countries: upstream of an already planned development process (*ex-ante*) or downstream of a spontaneous development process on which the government intervenes with a view of strategic organization of the urban area (*ex-post*). For example, cities of Roman origin (mainly located in the territories conquered in Europe outside Central and Southern Italy) see public planning as a 'fundamentally genetic' trait. In fact, these typologies of cities generally have their nucleus based in the organization of military camps. Within these military camps, soldiers often lived together with their families: so, around these camps, military-style social aggregates were developed organizing networks of economic, social, and commercial relations aimed at satisfying the inhabitants' needs. With the loss of military relevance of some of these 'camp towns - which was followed in some cases by the movement of military units to other areas of the Empire and in other cases by the transformation of the military units stationed there into groups of soldier-peasants - the civilian jobs have gradually replaced military jobs as a fundamental socioeconomic strain of the place. However, we must not forget that, in the experience of ancient Rome, the assignment to veterans of precise plots of land in conquered territories (*ager publicus*) responded, on the one hand, to the need of the Roman government to reward those people who had served in the army and, on the other hand, to proceed with the creation of Roman colonies within recently conquered areas, with a view to gradually assimilating the local conquered peoples to the typical lifestyles of the Roman civilization (Zanker, 2013). In other cases, the public planning action is located downstream of a development originated from a process of spontaneous occupation of the territory. This is the typical case of the cities of more ancient origins, including Rome itself, where the occupation of the seven hills near the Tiber river had created a urban agglomeration, to which the government will try only in a later period (Republican and Imperial times) to give an orderly structure by an organized management of urban spaces (Zaccaria Ruggiu, 1995).

Each management operation of the urbanized territory should be based on a precise evaluation of economic vocation and geographical peculiarities of the given place, as well as on the knowledge of the socioeconomic relationships characterizing local social life, especially when public action intervenes after a first spontaneous urbanization (Maliene et al., 2008). In particular, the objectives of public intervention require a planning process that takes into account the different possible alternatives (Zangrandi, 2019, p. 48). Therefore, definition of objectives follows the definition of plans and programs aimed at their achievement (Borgonovi, 1996), regarded as means preordered to the achievement of previously identified and calibrated purposes. The phases in which the business sciences separate the choice of plans and programs are essentially four: (a) identification of the areas of intervention; (b) definition of intervention alternatives; (c) effectiveness analysis of each alternative in relation to the specific objectives to achieve;

and (d) comparison between the different alternatives and choice of plans and programs considered most suitable and appropriate (Zangrandi, 2019). The business economics literature debated the distinction between concept of "planning" and "programming" (Borgonovi, 1996). There is wide support to the idea that planning is configured as "the management activity carried out according to given organizational modalities and with support of adequate information subsystems, aimed at (i) establishing the basic objectives (...); (ii) identifying the most appropriate strategic guidelines; and (iii) providing the technical and financial resources and skills needed to implement the strategies indicated in order to achieve the objectives set" (Airoldi et al., 1992). Programming is interpreted in business and economics sciences as "an activity which, on the basis of the strategic guidelines set out in the [planning] plan, identifies the concrete alternatives for action to be implemented in a short period of time that is generally identified with the calendar year, establishing the resources to be used and defining the tasks of the various managers" (Airoldi et al., 1992) and, consequently, the results expected by the various organizational positions involved. The relationship between the defined objectives and the plans (or programs) necessary to achieve them, call into question the concept of "rationality" of the public decision makers. In this regard, the economic literature identifies two possible models of rationality: "objective" rationality and "limited" rationality. The first model is typical of neoclassical economics and differs from the second model (typical of the managerial sciences) because it interprets social behaviors as non-habitual and non-routinized, as determined by the ability of rational calculation and progressive adjustments to the optimal margin. The concept of limited rationality is typical of managerial sciences (Simon, 1947, p. 143; Williamson, 1993, pp. 97–107), being based on a realistic approach to the relationship between objectives to achieve and the knowledge system in relation to the various possible alternatives. As Bonazzi (2008) points out "the neuro physiological limits of the human brain and, therefore, the limits of the human knowledge (. . .) are particularly evident when human beings (and enterprises) are confronted with complex realities and uncertain developments. The principle of limited rationality provides that men act as intentionally rational subjects, even if they are within boundaries that make their actions and predictions imperfect". This principle, as Williamson points out, has two important consequences. The first is the possibility to build a theory based on the assumption that the company rationally pursues efficiency goals. The second consequence is that "if the mind is a scarce resource (. . .) then an inevitable part in the economic research program is the study about the structures and procedures used to economize on limited rationality (. . .). In other words: becomes an integral part of economic research, the study of the choices among the most easily solutions controlled by a rationality that is known to be limited" (Bonazzi, 2008). The relationship between the choices of public decision-makers and the awareness of the unavoidable limited rationality of agents is fundamental to achieve good planning and good programming. In fact, as Capaldo (1965) effectively outlined, "for a good planning (...), it is not enough to set objectives; instead, it is necessary to seek - and with a great deal of adherence to reality - the ways through which it is possible to achieve them and, therefore, to make sure that, in the given time period, they are, within certain rationally set hypotheses, actually feasible. Otherwise the whole planning process becomes empty".

In the past, little attention had been deserved in explaining urban structures through analysis of the relationship of social classes within the city. In spontaneous cultures "where informality, community life and socializing, song and football attendance, mutual aid and illegal building are the main characteristics" (Leontidou, 1990), a specific approach should be adopted to analyze land-use patterns. With this assumption, Leontidou (1996) proposed a research method grounded on quantitative analysis considering social dynamics of settlements as the key element of change. This chapter takes advantage from Leontidou's human agency method of investigation in order to understand with more details how patterns of urbanization evolve and manifest themselves in Mediterranean Europe, a region heavily affected by recession.

The present study proposes a thorough analysis of the elements that determined postwar urban development in Southern Europe and argues on the necessity of considering the Mediterranean basin as an independent region requiring a specific investigation (Newman and Thornely, 1996; Zambon et al., 2018). In fact, defining a Mediterranean trend of urbanization presents some inevitable difficulties as the history of civilization in this region is particularly long and complex (Salvati 2018; Cecchini et al., 2019; Carlucci et al., 2019). While considering the complexity of this objective and the ultimate aim of this work, the subsequent section will directly zero in on the main spatial patterns which characterize the Mediterranean cities in the 20th century: the "Inverse-Burgess" model and the "vertical differentiation" typical of Southern European cities will be discussed (Brunori et al., 2018; Zambon et al., 2019a). Important elements that have played crucial roles in urban development of Mediterranean cities, such as the "popular land control" manifested through spontaneous building activities, will be also detailed in the following chapters.

This contribution will emphasize the need to study the intimate mechanisms of Mediterranean urban development by focusing on the social formation and modes of production of the region. In particular, special attention will be given to the subordinate economic forces (informal economy) and low social classes (poor people and workers) as these have been crucial actors influencing the dominant modes of production and social groups. Furthermore, the "adaptability" and "popular creativity" of Mediterranean subordinate social classes will be described and later connected to the urban asset of the city. The section concludes with a brief review of the development of Mediterranean societies through a human ecology approach, based on the observation of three basic dimensions: industrialization, urbanization, and 'the city as a material context'. The last part of the chapter debates on the concept of a (presumably homogeneous) Mediterranean urban ideology. How the people of this region live and perceive the spaces of the city, although with important local specificities, will contribute to understand and planning the urban patterns of (future) Mediterranean cities.

2.1 Spatial Organization of Mediterranean Cities and Distribution of Social Groups

There are many ways with which the Mediterranean region has been delimited. Some criteria are based on historical similarities between countries, whereas other criteria refer to climatic conditions or geographical characteristics (Francaviglia et al., 2019; Proietti et al., 2019; Salvati et al., 2019). For the present study, Braudel's boundaries of the Mediterranean region have been adopted. In "La Méditerranée et le Monde Méditerranéen à l'Epoque de Philippe II," Braudel defined the Mediterranean area according to a presumed homogeneity in agriculture, consumption habits, mental and cultural tools, language, religion, and laws. As a rule of thumb, the Mediterranean area has been defined as the "old world region where olive trees grow" (Braudel, 1953).

Another definition of the Mediterranean region has been proposed in the recent report of the European Commission "Europe 2000+." This definition is based on morphological characteristics of the region and divides the Mediterranean region in six "environmental frameworks": the Latin arch, the Adriatic basin, the Maghreb front, the Libian–Egyptian area, the Middle-Eastern façade, and the Balkan bridge.

The Mediterranean is a "sea in between lands": its region is delimited by the three continents of Europe, Asia, and Africa. Due to its geographical characteristics, it has always been at the center of commercial and cultural exchanges. Throughout history, conflicts between different populations for the domain over the region have been frequent and intense. As a result of the various collisions between different civilizations, the Mediterranean region is extremely heterogeneous and unique (Cuadrado Ciuraneta et al., 2017). The heterogeneity of the Mediterranean region makes it impossible to construct a single model of urbanization for the whole area (Carlucci et al., 2017). The cities of the Mediterranean have experienced diverging patterns of urbanization throughout their history. According to Spina (2004), there are four dominant urban typologies in the Mediterranean: (i) the Roman city, (ii) the Hellenistic city, (iii) the Byzantine city, and (iv) the Arab city.

Mediterranean urban histories in these countries have presented a "wealth of the most bizarre combinations" (Gramsci, 1917). Even if originating from the same capitalistic societal formation that characterizes the Western world and Northern Europe, Mediterranean countries have (and are) experiencing diverging trajectories of urban development. The spontaneous urban development through popular land colonization has been a common phenomenon in Mediterranean cities. In this line of thinking, the working hypothesis is that spontaneous urban expansion is neither a precapitalist remnant, nor a manifestation of residual peasant modes of land allocation.

It is believed that urban spontaneity has emerged with capitalistic development and has been "functional" to it. If defining the Mediterranean region may have seemed complicated, then just imagine the difficulties that historians and geographers have encountered in trying to construct a Mediterranean city archetype. In the various definitions that have been proposed for the Mediterranean city, there are few common features and an abundant list of particular characteristics. This vast quantity of divergent features between the cities of the same region makes it practically impossible to construct an archetype of the Mediterranean city. For this reason, it is convenient to abandon the search for a unique Mediterranean city model in favor of a series of urban typologies that are suited to consider the different geographic, economic, social, political, and cultural characteristics.

The Mediterranean cities seem to share many characteristics but, at the same time, they present a series of inconsistencies that make it practically impossible to analyze Mediterranean urbanities with the use of a single city model for the whole region. While some authors believe that the limitations for studying the Mediterranean region as a whole rely in the diverging characteristics of its cities, other scholars are attracted and fascinated by such diversities. The originality of the Mediterranean, according to Braudel (1953), can only be correctly explained and described through the interaction of biophysical and anthropological factors. The peculiarity of this region, deriving from a long history of population conflicts, dialogues, and exchanges, requires a specific analysis rejecting linear logic and statistics in order to fully understand its complex nature. The conclusion is that Mediterranean spaces cannot be "weak" or "inconsistent" because it is difficult to define when adopting a binary spatial logic (inclusion/exclusion) and linear interpretation of cartography. "Quantitative" approaches will inevitably lead to a "reduced" and "marginalized" understanding of the Mediterranean spaces. For this reason, "qualitative" approaches based on human spatial relations, are particularly appropriate to assess urban dynamics in the Mediterranean region (Chelli and Rosti, 2002; Chelli et al., 2009; Rosti and Chelli, 2009).

The works by Albert Camus on Algeria, the detailed descriptions of Southern Italy by Antonio Gramsci, and the work of Franco Cassano inform further research on the characteristics of Mediterranean cities and the prevailing urban culture of this region. According to Camus, the Mediterranean tradition is characterized by a "solar ideal" manifested in the society by a reflexive liberty, an individualistic altruism that privileges elements such as limits and nature over excess and history. The French Algerian author has dedicated a part of his literary career in describing "le pensée de midi" (the mentality of the South), which he believes to be a positive and solar way of being. According to him, the Mediterranean philosophy of life is based on the "juste mesure" (literally meaning "the correct measure") of human beings. In this sense, Mediterranean cities do not give that perception of a dehumanized space. Camus positively emphasizes how the natural qualities of the Mediterranean are kept alive within the city. In explaining Camus' love for the natural environment of the region, Jacques Chabot said that the French author's philosophy originates "more from the Mediterranean beaches than its libraries."

Camus emphasizes the concepts of the "solar ideal," the harmony between man and nature, the family stands out as the basic element of Mediterranean societies from Gramsci's analysis on southern Italy. This characteristic may result functional for explaining the feeble urbanization of the Mediterranean, where the well-being of its inhabitants is often weak (Corona et al., 2016). In this sense, urban planning seems to be overwhelmed by a familiar spontaneity, reinforced by an attitude of informality and illegality, which creates a strong cohesion between the members of a society (Leontidou, 1996). In other words, the spatial organization of Mediterranean cities and the distribution of social groups are influenced by the strong relationship between relatives. Franco Cassano, in his work, entitled "Il Pensiero Meridiano" (The Meridian Thought), highlights another important characteristic of the Mediterranean societies that results useful for understanding the urban development of the region. According to the Italian sociologist, the respect of rules, typical of Central and Western European societies, is contradicted in the Mediterranean by a strong attitude toward "dispensation." Cassano believes that a society in which the permission to ignore a rule or an obligation is only possible within a social context characterized by a tradition of honor and a way of being that favors a peaceful harmony with nature rather than the "ethic power of the State" (*sensu* Hegel). This tradition is typical of a Mediterranean region regarded as a place where the alliance between man and nature derives from the Greek myths and divinities. Renouncing such tradition and breaking the harmony between human beings and the natural environment might result in making these regions become an "erroneous copy of the North."

2.2 Economic and Political Structure of Mediterranean Societies in Relation to the Development of Urban Spaces

The economic and political structures of a nation play a central role in the development of urban spaces. According to Marx and Engels, the form and functions of the city are direct consequences of socio-economic organization. Both the relations between different cities and living conditions within a specific location can be explained as responses to the mode of production of a given historical moment. Despite each Mediterranean city has its own specific features, the traces of thousands of years of economic and cultural exchange, interdependence and successive domination can still be seen today within the various Mediterranean urban geographies, in their physical and social organization as well as in their most everyday operations.

The Mediterranean region, as part of Europe, saw capitalism gradually develop as a system from the 16th century. The main characteristic of this economic system relies in the nonlabor factors of production, such as capital and land. Even if it is not easy to find a unique definition, most economists agree that prices, wages, and the private ownership of the means of production and creation of goods or services for profit in a market are the main elements of capitalism. In the 19th and 20th centuries, capitalism became dominant in the Western world and provided the main means of industrialization throughout much of this area. Again, urban settlements played a key role in the emergence of capitalism but, at the same time, they were strongly shaped by its needs. In the history of capitalism, space has always developed through stratification (Leontidou, 1990) and the working class confined to residual land, lacking in basic urban services. However, this is the only universal statement that can be made about capitalist cities. In order to understand how spontaneous urban development patterns have originated throughout the Mediterranean, an analysis based on the social forces that have taken part in the process shall occur (Di Feliciantonio et al., 2018).

The Mediterranean region contains some of the oldest cities of Europe, many dating from ancient time. Their history is much older than many Northern European cities that have developed only after the end of the Middle Age. This means that urban patterns of Mediterranean cities have developed over a large time, experiencing a more complex succession of governments, economic systems, and social organizations (Salvati, 2018; Carlucci et al., 2019; Cecchini et al., 2019). Overpopulated agglomerations of the past determined the shapes of today's modern cities. Beginning with Caesar's Rome which contained inside its walls (delimiting an area of 15 square miles) approximately 400,000 inhabitants, the Mediterranean region hosted some of the first urban monsters. Athens is considered as the dominant city in classical times and Constantinople the glory of Byzantium. Until the 19th century, when London and Paris emerged as big cities, Istanbul and Naples were the most densely populated cities in Europe.

After the mid-19th century, modern metropolis arose when the walls of the old cities were pulled down. The industrial revolution, the demographic boom, globalization, and the transformation of the economy all played critical roles in the urbanization process of cities (Monarca et al., 2009; Clemente et al., 2018; Ciommi et al., 2018; Salvia et al., 2018; Zambon et al., 2018; Carlucci et al., 2019; Cecchini et al., 2019a, 2019b; Moretti et al., 2019; Salvati et al., 2019a, 2019b, 2019c; Dimski, 2005; Zambon et al., 2019a, 2019b). Since World War II, Mediterranean and European countries have experienced similar models of urbanization, resulting in similar structures and population levels. "Overurbanization" and real estate speculation were the general trend. Most of the responsibility for such patterns of growth can be attributed to the demographic growth of the region: by 1971 Rome, Barcelona, and Athens had between 2.6 and 3.5 million inhabitants (Figures 2.1–2.3) with annual

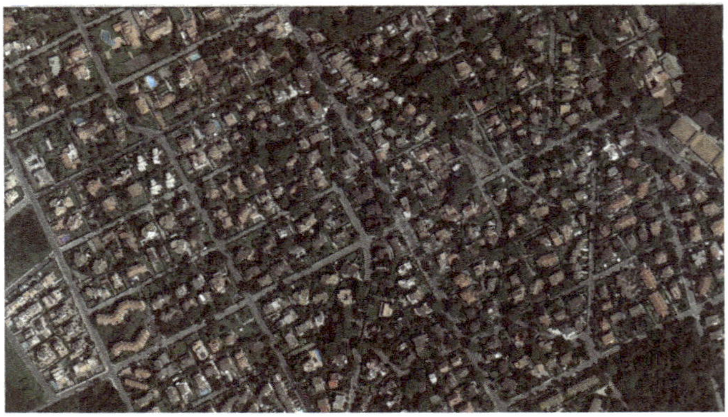

Figure 2.1 Suburban Barcelona. *Source:* Google Earth.

Figure 2.2 Suburban Athens. *Source:* Google Earth.

growing rates ranging from 2.8% to 3.3% (Ferras, 1977; Leontidou, 1990; Costa, 1991).

Another important factor that took part in the exponential and rapid growth of Mediterranean cities during the 20th century was immigration. From the coasts of Northern Africa and Eastern Europe, most of the Mediterranean region experienced intense immigration. This phenomenon has had relevant consequences on the social organization of these cities. As Italy, Spain, Turkey, Greece, and Portugal are becoming more developed, the native-born population is dropping rapidly and at the same time the presence of foreign immigrants is increasing. As a result, "Rome, Madrid and Athens are demographically all new cities" (Fried, 1973).

Figure 2.3 Suburban Rome. *Source:* Google Earth.

2.2.1 Models of Urban Development Organization in the Mediterranean Socio-economic Context

In 1925, Burgess presented a descriptive land-use model, which divided cities in a set of concentric circles expanding from downtown to the suburbs. This representation of the urban development organization was built from Burgess' observations of several American cities, notably Chicago, for which he provided empirical evidence. The model assumes a relationship between the socio-economic status of households and the distance from the inner city. The further from the core city, the better is the quality of housing, but longer the commuting time. Thus, accessing better housing is done at the expense of longer commuting times (and costs). The exodus of middle-income families and the social elite that gave birth to suburbia in Northern Europe is contradicted from the path of urban growth in peripheral capitalist countries (Ciommi et al., 2018, 2019; Salvati et al., 2018). The dominant trend observed in the Mediterranean cities consists in the reverse pattern: the social elite inhabiting the center and the poor families displaced in spontaneous built up areas. This spatial pattern has been defined as inverse-Burgess (Leontidou, 1990).

The inverse-Burgess spatial pattern characterizes large metropolitan cities throughout the Mediterranean Europe. It describes the distribution of social classes over the territory according to land rent and density gradient values. In contrast with the Northern European urban trend, in the Mediterranean cities there was a higher demand and desirability for city center dwelling units. As a consequence, the rich bourgeoisie tend to cluster in the center of the city, where land-rent values and population density are high, while the working class move toward peripheral areas where the costs of life are affordable (Colantoni et al., 2018). For instance, social rank in Italy has been found to be low in the Roman suburbs (and still lower in the Agro Romano on the urban fringe of Rome) and particularly high in the old city center (Krumholtz, 1992). Also, in Madrid and Lisbon central areas and inner 19th century suburbs had the highest socio-economic status (Costa et al., 1991). The inverse-Burgess tendency is one of the main reasons that brought gentrification in the Mediterranean cities (Carlucci et al., 2018). As desirability for central areas grows, many of these districts have undergone intense requalification programs and interventions. Central historical slums, typical of old towns, have been strongly requalified and thus risen in value. Therefore, native inhabitants of these central slums have been "forced" to move to peripheral areas by market dynamics, leaving central spaces to high-income social classes (Seronde Babonaux, 1983). Some examples of this process can be observed in the central districts of Rome, Lisbon, and Barcelona (e.g. Fratini, 2000). Even if the inverse-Burgess model is the dominant trend in Mediterranean cities, this does not mean that spatial patterns of urbanization always follow it precisely.

Some events or political decisions may bring variations to the normal urban trend. For example, "garden-suburbs" have sometimes attracted bourgeoisie and middle-class families to abandon the center of a city. But the zoning tendency in the Mediterranean cities is far from reaching the success that it had in Northern Europe. Furthermore, the inverse-Burgess model does not imply that the centers of Mediterranean cities are inhabited exclusively by middle- and high-income individuals. Community segregation, typical of Northern European cities, is generally pervasive in the Mediterranean cities which favor a social mixture in the center. Nevertheless, there is a "spatial separation" between the middle and working class, but not on a horizontal plane. Except for some slum areas and modern housing districts, the middle and working class live together in vertically stratified apartment blocks: service laborers and working class in lower floors and wealthier workers and renters on top levels in penthouses (White, 1984: p. 156). This social separation on levels of height has been called "vertical differentiation." The central areas of Naples represent a manifestation of this phenomenon, especially because the city has few peripheral shanty towns (Allum, 1973).

Vertical differentiation can be largely attributed to the proliferation of the multistorey apartment building (Brunori et al., 2018; Zambon et al., 2019), which apparently originated in the Italian cities of the Renaissance and gradually spread to Europe in a complicated historical process starting in the 16th century in France and Austria. In the 20th century, the intensification of building height and the proliferation of multistorey apartment blocks in Southern Europe were the consequences of housing shortage (originated from the economic recession of the 1930s and World War II) and rising prices of land and buildings. This phenomenon has produced a compact and overcrowded conformation of the Mediterranean cities. For instance, green space declined from 2.8 m^2 per capita in 1925 to 1.5 m^2 in 1964 in downtown Rome (Fratini, 2000). The location of the economic activities provides us with another contrast between northern Europe and the Mediterranean spatial patterns of urbanization. In northern cities, a neat organization of economic activities in specialized zones located according to the land-rent gradient is the result of a process of economic rationalization and planning. Southern cities, by contrast, are characterized by a disordered urban tissue and a patchwork of economic activity. Retail and artisan establishments, kiosks, and workshops of the informal economy are scattered in small local centers, along roads and in residential areas (Leontidou, 1990). In other words, zoning is rare in the Mediterranean cities, while mixed land uses seem to be the general trend.

Furthermore, the late industrialization of the Mediterranean region evidences the absence of "bourgeoisie hegemony," but above all of heterogeneity and diversity. In the Mediterranean cities, the subordinate social classes have often revealed the process of urban development as the main driving force. Not necessarily in opposition with the dominant classes, workers and poor individuals have strongly influenced the popular tradition of these places. Consequently, an urban reality characterized by the cohabitation of low- and high-class social groups in the same space has developed throughout time (Balchin, 1996). The urban tissue of the Mediterranean cities is chaotic, heterogeneous, with housing areas hosting multiple activities and vertical differentiation being the norm (Salvati et al., 2013; Brunori et al., 2018; Zambon et al., 2019).

All this, together with an informal organization of labor activities and inefficient planning systems, has brought to the overlap of working and residential activities in the same spaces and to the dispersion of employment throughout the entire city. For these reasons and particular features of the Mediterranean societies, cities of this region are usually very compact, densely inhabited, and lacking functional differentiations. There are references of restrictions against land invasion and effective controls over illegal building activities from a very early period in the cities of the north. Even during the period of transition to capitalism, when urbanization rates were very high, squatting attempts by poor social classes in Northern European cities have been limited and every kind of illegal building promptly suppressed. Popular land control was out of the question in Anglo-American cities of the north.

The control of peripheral urban land seems to have been much looser in the Mediterranean, especially Greece, since the 19th century (Ioannidis et al., 2009). Semi-squatting was a diffused practice that gave birth to several shanty towns in the peripheries of the Mediterranean large cities. To the south and east of the Mediterranean, the strength of the so-called "informal" housing was dominant. Depending on the country and the conurbation, between 30% and 70% of town dwellers are only able to construct their own homes by working through informal channels. Real estate speculation and the lack of planning are considered as the general traits in postwar Mediterranean urban development. Urban expansion has occurred mainly though illegal and speculative initiatives (Zambon et al., 2018). The illegal building sector flourished in Southern Europe during the 19th century through popular initiative, even if there are some references of illegal dwellings built with subsidies from the State (Fried, 1973). In peripheral areas, entire illegal buildings were erected while in the inner city a standard speculative action was the construction of illegal rooftop additions within the approved urban plan, which gave the middle classes supplementary dwellings for exploitation.

Mediterranean large agglomerations were thus built bit by bit, and their physical expansion was defined by where people erected their precarious settlements. Italian cities, besides some exceptions in the north of the country, grew mainly through popular land colonization. For instance, Rome evidences a clear pattern of spontaneous development, with approximately 200,000 people reported as living outside the plan perimeter of the city, in scattered *borghetti* (self-built shanty towns along roads illegally divided into allotments), *borgate ufficiali* (areas built under the official auspices of the fascist government for those expelled from the city during the reconstruction of central areas), and *borgate* (self-built illegal settlements on the urban periphery). Spanish large cities are frequently surrounded by *viviendas marginales*. Official census estimates shanty-towns households at 1.1% of the total households in Madrid, 1.4% in Malaga, and 2.6% in Seville. Barcelona had most of its proletarian housing (most of which illegal) built in the 1920s when the city attracted Andalusian laborers for the construction of the metro and the Barcelona Fair.

The squatting problem was even more intense in Portugal. Shanties in the *bairros clandestinos* (or *barrios de la lata* as they are known locally) are as numerous as the Greek *afthereta*, but they are much more destitute and constructed with perishable materials such as cardboard, wood, and corrugated iron. According to official estimates, 6.4% of Lisbon's total population was housed illegally. In the case of Athens, during the postwar period, colonization of peripheral zones by the proletariat and floods of migrants to central areas has been observed. The core of the semi-squatting movement which colonized peripheral areas was guided by the proletariat employed in the industrial and construction sectors. On the eastern edge of the Mediterranean region, Istanbul still expands through *gececondus* (settlements built overnight). The once glorious city of the Levant, together with Ankara, contains most of the Turkish squatter population. In 1980, 23.4% of the total urban population of Turkey still lived in *gececondus*.

2.3 Economic and Social Determinants of Local Development Process: The Mediterranean Experience

Spatial distribution of social classes alone is not enough for developing a theory of urban growth in the Mediterranean countries. After all, every city in Europe and the United States traditionally has central areas inhabited by wealthy individuals, pockets of poverty in the inner city, some wealthy suburbs, and low-income working class concentrated in peripheral areas. The risk is that "an urban growth theory structured upon spatial distribution becomes vulnerable to criticism." For this reason, in order to fully understand the urban development of the Mediterranean cities, social relations will have to be considered as dominant variables of the analysis. According to Leontidou (1990), spatial distributions and tenure patterns should be considered as the visible and quantifiable outcomes of the interactions between social classes and space: "what actually differentiates urban development patterns in capitalist cities is not spatial distribution but land control according to social class." Informal modes of allocation in the Mediterranean cities have produced a variety of adaptations, tenure categories, location, and density patterns. Economic and social land allocations are the two ideal-type models used for studying urban transformation. The former investigates spatial distribution of settlements as a consequence of economic forces while the latter analyses the social relations existing between different classes. In real life, many modes of land and housing allocation can coexist in the same capitalist city, thus proving the impossibility of conducting an urban analysis based on these two ideal types. In fact, dominant and subordinate modes of land allocation have coexisted with various intensities in different urban systems.

In order to conceptualize a more developed model of urban growth, the social formation and modes of production of a city must be considered (Kazemzadeh-Zow et al., 2017). On the one hand, modes of production in the urban sphere are useful for constructing urban types. But as cities are generally characterized by a mixture of economic systems, these urban types remain valuable only for classification purposes. On the other hand, social formations can be more directly related to urban formations, that is, articulations of economic, material, political, and social forces from various urban types. For emphasizing the importance of social formations, let us give a look at how paradoxes derived from studying spatial distribution only through modes of production can be solved.

Market is the dominant mode of land allocation in capitalist urban formations. According to traditional urban theory, uniformity of land allocation should be the result of competition between social classes based on rent-paying ability. Analyzing spatial distributions through the dominant mode of production would result in using bid-rent curves as the main variable. As a consequence, it would be possible to find the following spatial pattern in capitalist cities: concentration of the high- and middle social classes in the center of the city, where land per unit area is more expensive, and the low social class allocated in peripheral areas where land costs are lower.

The Anglo-American reality is in contrast with this predicted spatial arrangement of social classes. In cities of Northern Europe and of the United States, high- and middle-income social groups concentrate in suburbs, while working classes tend to concentrate in central locations on more expensive land. This is a clear paradox. But if other variables would be considered in the analysis, such as costs of commuting, urban conditions, neighborhood, and dwelling typology preferences, then the paradox can be solved. Wealthy population segments in Anglo-American cities prefer to live in new spacious expensive villas located in healthy areas, leaving the old and small apartments of the crowded center to the low- and middle social classes. Furthermore, commuting costs have a strong impact on the poor and thus represent another variable that forces low social classes away from suburbs.

This paradox is not shared by cities of the Third World, where the social structure presents some evident differences with the European one. In "peripheral cities," "a high degree of correlation between the price of land and the socio-economic status of the class residing on it can be observed" (Amato, 1970). This can be explained by looking at the social formation of the Third World countries. The high-income social class represents a small portion of the population and the gap with the other social classes is much larger Western than the one of other societies. Therefore, the social elite can allow buying buildings in the center of the city, without having to search for spacious residential units in the suburbs.

The modes of production and the social structure of the Mediterranean cities are very similar to the Anglo-American type, but urban patterns seem to contradict this similarity. Socio-spatial patterns of distribution in the Mediterranean region are characterized by concentration of wealthy households in expensive large downtown apartments while the poor social classes are forced to move to outer peripheral areas. In order to explain this pattern, it is useful to focus on the related socio-cultural forces. Alongside the market, the dominant mode of production in determining land allocation usually outlines subordinate modes that operate within the formal borders. Their emergence is not a departure or distortion of capitalism, but one of its inherent characteristics necessary to solve contradictions created by the dominant mode of land and housing allocation (Leontidou, 1996). Subordinate modes of production compete and combine with other modes present in the same societal formation at different scales (economic, social, political, cultural, and ideological).

Urban patterns have been usually studied by considering the "dominant" classes of society, that is, landowners and people who can choose their residential space without any relevant material constraint. By contrast, in the Mediterranean context it is more convenient to focus on the working class, a social group that has hardly ever considered along the lines of a single mode of land and housing allocation and that represents a large part of society. Location and tenure patterns of the working class are constrained by a series of forces, which result in the emergence of a variety of housing conditions. Such variations make the proletariat perhaps the most crucial class in any study of urban development. But finding a theory that successfully explains working class land allocation patterns is complicated, as this is influenced by an articulation of forces comprising the urban economy and society, the system of production, distribution, and consumption.

2.3.1 Subordinate Classes and Informal Economy

In the attempt of finding an explanation to spatial patterns of the Mediterranean cities, the analysis of composite behaviors and actions of the working class reveals an appropriate point. Three dominant characteristics of this social group have been observed: (i) adaptation (i.e. the ways with which classes adapt within dominant mode of land allocation),

(ii) popular creativity (i.e. the logic of emergence of subordinate modes of production), and (iii) spontaneity/informality. These represent the key features for understanding working class land allocation patterns in southern Europe.

Informality is an unofficial attitude which permeates many spheres of urban life, peripheral and semi-peripheral capitalist cities. Urban growth of southern European cities has occurred in the 20th century mainly through the creation of suburban communities by popular initiative and creativity (Ciommi et al., 2018, 2019; Salvati et al., 2018). The frontiers of urban expansion are determined by popular land colonization and illegal building. In the Mediterranean region, social, cultural, and economic forces have brought the proletariat outside the "official norms of the middle-class city." This mode of land allocation is less frequent in advanced capitalist cities, where the dominant classes have determined the process of urban growth in a totally different way. As a result of these three characteristics of subordinate classes, peripheral informal settlements are widespread and systematic features of the Mediterranean cities: *viviendas marginales* in Spain; *gecekondus* in Turkey; *bairros clandestinos* in Portugal; *borgate* and *borghetti* in Italy; and *afthereta* in Greece. These have been usually interpreted as the consequence of housing crisis. However, this crisis has not created such urban developments in northern and central Europe, so there must be other causes for the abundance of peripheral informal settlements in the Mediterranean region (Salvati et al., 2016). Looking at the social formation of southern European countries, a general informal attitude and tolerance for illegal settlements and land occupation can be noticed. This could probably explain why unauthorized peripheral suburbs have developed in the Mediterranean. Illegal settlements refer to unauthorized use of land, not illegal occupation: plots of land specified as "agricultural" by urban plans have been sometimes used for residential purposes. Semi-squatting practices had also been a relevant process in the spatial distribution of population in urban areas. Illegal occupation of property and land by individuals, a common practice in the Third World countries, has characterized peripheral areas of southern European cities during the first years of the 20th century. As economic development, together with rising incomes, has progressed in the Mediterranean region, squatting has given way to "semi-squatting" (Abrams, 1964), a milder form of the phenomenon. Semi-squatting is a subordinate mode of land and housing allocation to the dominant market yet creating its own operating rules.

The informal sector, a concept introduced by Hart (1971), reveals to be useful in understanding peripheral and the Mediterranean capitalist social formations. Informal economy is considered as an integral part of the capitalist systems, its functioning and development, as well as the structure of peripheries. But some attention should be placed in defining which sector of informal economy can be considered for the investigation of spatial development, as this sector includes "black economy," "para-economy," illegality, usury, and crime. Our study will refer to the "lighter" side of informal economy, which is related to urban development. In the Mediterranean region, the "lighter" informal economy is represented by free laborers, putting-out workers, self-employed artisans and shopkeepers, family entrepreneurs, and pedlars (Leontidou, 1990). Large and modern enterprises of the formal sector and a constantly growing small-scale sector of shops, artisans, and "free" laborers of the informal economy produce a polarized society. The socio-spatial subdivision between "giant" and "dwarf" enterprises is typical of the Mediterranean cities and has been studied for instance in Naples by Allum (1973).

"Not legal economy" is therefore one of the key elements of the Mediterranean region that still today has not lost its importance. Even if informal economy is a typical characteristic of developing countries, the phenomenon presents different traits in the Mediterranean basin. Above all the Mediterranean spontaneity results in recognition of the role of capitalistic market, refusing state as the regulator of market dynamics at the same time. Furthermore, the role of informal economy in the Mediterranean is not anymore connected to the necessity of surviving, guaranteed by an acceptable level of development, but to the "social promotion" of subordinate classes. In this view, informal economy should be interpreted as a feedback mechanism to the inefficiency of the market and to the fiscal "avidity" of the state, unable to contrast unemployment of young people, as well as social and infrastructural gaps. However, the economic spontaneity and "autopromotion" are typical characteristics of the Mediterranean societies.

2.4 Human Ecology and Southern Cities

Human ecology promotes a materialist urban sociology, since it stresses spatial-economic forces of social groups. Moreover, it considers relationships between social classes and the city in an ecological perspective, where adaptation, segregation, filtering, invasion, and succession are the main observed variables (Ciaccia et al., 2019). Various ecological complexes can be conceptually constructed for analyzing urban structure and growth, social organization and collective adaptation. Research on the Mediterranean urban patterns carried out by Leontidou (1990) has revealed that spatial analysis does not support the formalism of ecological complexity. Three basic dimensions have shaped the spatial distributions of settlements in the Mediterranean region: industrialization, urbanization, and "the city as a material context." Each of these elements became predominant in certain historical periods, keeping the others subordinate, imposing structural principles and explaining urban spatial patterns and transformations. For example, during the interwar period, urbanization was predominant due to refugee arrival while, during the postwar period industrial restructuring and the city as a material context (the expansion of material capitalism and environmental deterioration) were determinant in subsequent urban processes. Now, we present a detailed explanation of these elements.

2.4.1 Economy, Industrialization, and Urbanization

The relationship between urbanization and industrialization is radically changing as capital becomes more mobile. The internationalization of socioeconomic relationships in the 1970s meant that the United States and the European companies could locate their production plants overseas, to low-wage countries, transforming the structure of peripheral world, and causing the emergence of new industrializing countries. Furthermore, this process was scaled up together with the division of the production process in three levels: conception and engineering; qualified manufacturing requiring skilled workers; and deskilled execution and assembly (Wallerstein, 1979). Despite important linkages between population growth and industrial concentration, centripetal urbanization was not created by industrialization. A process known as "urbanization without industrialization" affected peripheral economies (Leontidou, 1990). Cities affected by this process later caused relevant floods of migrants in areas of potential employment. But the rapidly increasing labor force was not in the formal capitalist sector. The informal, low-productive, unprotected, and unstable employment network grew with the industrialization process of peripheral capitalist cities. By contrast, "core" capitalist cities of the 19th century were shaped through a strong connection between industrialization and urbanization. This is due to the fact that technological change and the market made the early capitalist cities "self-contained employment areas" (Robson, 1969). As labor became a commodity to be bought and sold in the market, States have often treated housing and land allocation as a means of manipulation of the labor market in the interests of industrial capital.

Mediterranean Europe stands between the two contrasting social models of "core" and "peripheral" cities. From a conceptual point of view, it can be said that southern European societies belong to the semi-peripheral world. For instance, Italy has not been considered in this class of world societies, as it was one of the most developed countries of the Mediterranean and amongst the first members of the European Economic Community. However, the Italian Mezzogiorno is still considered an underdeveloped area in southern Europe. Despite the rise of capitalism through the concentration of business firms as early as the 16thcentury (Braudel, 1953), modern industry arrived late in the Mediterranean and had a weaker influence on urbanization. As a consequence, urbanization of the Mediterranean cities has been discussed with little reference to industry. In this region, industry followed rather than created urban concentrations, in sharp contrast to the case of northern and central Europe. "To some economists, Rome is a classic study of urbanization without industrialization," and the Italian capital is not the only example for southern Europe (Fried, 1973).

Even if economic development speeded up during the aftermaths of World War II, countries in southern Europe remained the "proletarian nations" of the continent, exporting their surplus labor to the north. This emigration from the Mediterranean region to more advanced countries played a crucial role in the urbanization process of southern Europe during the 20th century. But labor emigration floods have been slowed down since solid economic development occurred in the Mediterranean region after World War II. The "Italian economic miracle" began in the 1950s in the north of the country, followed by the Greek and Portuguese miracle in the 1960s. Spain's development was delayed due to the presence of dictator Franco who isolated the country from the rest of Europe.

2.4.2 City as "Material Context"

The city as a material context consists of the natural and built environment which transforms indirectly because of the technological and organizational change, related with cultural and institutional aspects of the society. This transformation was indirect because technological and organizational aspects of a society first determine the exact nature of land market which later defines the urban structure and growth of a city, independently from industrialization and urbanization processes. Technological aspects of the built environment include housing and utility production in the city, and transport, the "technological destruction of distance" (Handlin, 1963). None of them are purely technological, since even transport requires organization for its widespread application. However, they are connected with the material development that affects urban restructuring: "urban transformations are a consequence of technological development and particularly improvements in communications" (Robson, 1973). For example, technological aspects of the real estate market, such as type, availability, and costs of production of housing and infrastructure (i.e. land, building materials, and labor) affect rates of building and relevant shortages. However, organizational aspects, such as the structure of competition within planning systems and urban landownership, are often more crucial. These two variables of the material context of a city play a significant role in urban restructuring. Therefore, understanding their nature represents an opportunity for analyzing urbanization dynamics.

The material context of a city transforms through technological and organizational changes, but these are not the only variables which take place in the process. The role of the natural physical environment of an area should not be underestimated while analyzing urban development. Climatic conditions play a role in urban landscapes and lifestyles of a city: "At the heart of this human unit (. . .) there should be a source of physical unity, a climate, which has imposed its uniformity on both landscape and ways of life" (Francaviglia et al., 2019; Proietti et al., 2019; Salvati et al., 2019).

2.5 Mediterranean Urban Ideology as Cultural Basis for Management of Urban Development

Urban development is a product of the social formation of a city. In this section, the necessity of adopting the "human agency" as the main method of investigation has been discussed. The transition between models of urban development can be explained as an outcome of both transformations of material elements of the ecological complex and social class struggles and behaviors (Salvati 2018; Carlucci et al., 2019; Cecchini et al., 2019). From the study carried out by Leontidou (1996), it is evident that urban development patterns of southern European countries present a reversal trend from the one of the most advanced societies of the continent. The reversal was related to the structural development of the cities, to the process with which transformations occurred, and to the cultural orientation toward urbanism. While discussing over which methods and variables should be adopted for analyzing urban systems, the importance of human agency, social competition, urban cultures, and the historical development of class relations have been stressed.

By using these elements and approaches, certain aspects of urban cultures can be tackled effectively, and to a great extent help explaining contrasts in urban development trajectories between core and peripheral societies. For this reason, these variables should be considered as endogenous for every model of urban development, thus creating a model of social action rather than a behavioral geography. In other words, cultural traditions, meanings, collective practices, and their transformations will prevail as key variables for analyzing metropolitan development and regional sustainability (e.g. Chelli et al., 2016; Gigliarano and Chelli, 2016; Ciommi et al., 2017).

Between northern and Mediterranean Europe, there are contrasting orientations toward urbanism and many other aspects. Anglo-American dominant classes have traditionally idealized nature and tried to live next to the countryside (Robson, 1969). The social gentry-built country cottages safeguarded peripheral urban land very strictly. Consequently, besides controlling the center of the city (for economic purposes), the dominant bourgeoisie raised and protected, for their interests, peripheral land (for residential purposes). Thus, it is possible to conclude that northern cultures are characterized by an antiurbanism ideology, in which the countryside is exalted and the flight from the city has been a constant theme in the intellectual tradition (Wynn, 1984). The very antithesis of the above cultural orientations to urbanism is found in the Mediterranean. In southern countries, a prourban ideology can be observed since very early times, which identifies the city with progress and civilization, while the countryside was considered as the domain of ignorant peasants (Couch et al., 2007). Furthermore, due to the warm climate, Mediterranean populations are encouraged to outdoor life in the city, transforming open public space in places for social exchange and community life (Salvati et al., 2019). Suburbanization emerged in the 20th century coinciding with a progressive change in the Mediterranean urban cultures. As congestion and environmental concerns emerged since the 1960s, the idea of "suburbanism as a way of life" was growing in the middle classes (Ciommi et al., 2019).

Antiurbanism has always peaked during periods of authoritarian regimes in southern Europe. In Franco's Spain, an antiurban ideology was preached by State-controlled institutions: the city was seen as the center of vice and evil (communism, divorce, prostitution, and crime). During the Fascist era, the authoritarian regime stressed the importance of rural development, and the "peasant Italy" was idealized as an alternative to the ills of urban living. Furthermore, antiurbanism laws and legislation on forced domicile were attempts of the authoritarian regime for controlling rural–urban migration which peaked in 1936. In Athens, a set of policies aimed at limiting the growth of the city through the control of popular house building and the facilitation of industrial decentralization were passed. The antiurbanism ideology produced by authoritarian regimes and the growing problems related to high density (i.e. urban congestion and environmental pollution) in the city centers have constituted the undercurrent of middle class suburbanization waves (Carlucci et al., 2017).

2.5.1 Urban Expansion and the City Idyll

Urban expansion thus began with the first urban exodus of the social elite beyond the city limits because of their desire for "pastoral lifestyles" near the countryside. Satellite suburbs were created with the aim to escape from the congested, chaotic, and crowded urban areas in search for some peace and relax in a rural idyll. Later in the years, urban policies were shaped upon these geographical imaginations. Green belts, health legislation, safeguarded suburbs, gated communities, and tightly controlled land use regulations in the countryside were the modalities that accompanied city development in most of northern Europe and North America (Hall, 1997). The Anglo-American urban cultures that determined urban expansion are in contrast with the ones that characterized the Mediterranean cities. Urban residents of southern Europe have not been seeking "rural utopia." In fact, from the early years of the 20th century, positive geographical imaginations for urban life ("urban idyll") made the city appear as a shield and an antidote from rural poverty and insecurity. In the Mediterranean Europe, urbanity was synonymous of economic prosperity, better job opportunities, and social amenities or infrastructure linked to a higher quality of life. This culture for urbanism (in Greek "Astyphilia") was one of the main causes for growing rates of migration from rural areas to the city core. Consequently, there was a strong demand for urbanity in the Mediterranean societies.

"Rural exodus" provided cities with new residents, generally poor people of low social status, in need of cheap accommodations that would give them fast access to the economic opportunities located within the city's boundaries (e.g. Rosti and Chelli, 2012). Taking advantage of weak planning systems and political fragility of most governments during the interwar period, a long history of spontaneous popular suburbs and illegal sprawl began, especially in peripheral areas close to developed infrastructure. Urban expansion was thus caused by the urban poor who wanted to approach the city in the cheapest possible way. This urban development is particular and therefore cannot be analyzed using the general theories, as the "Alonso-type" models of urban land use based on rentpaying ability, land rent/distance curves and the rational behavior of urban actors, or "life-cycle" models of urbanization-suburbanization (Morelli et al., 2014). Spontaneous popular suburbs were made possible mainly because governments turned a blind eye on the phenomenon, initiated by popular strata. Illegal settlements were appearing all around the city center, thus representing a "compact" form of sprawl. These were usually the product of speculative entrepreneurs and landlord actions that subdivided large properties and later sold them to migrants who started building in violation of the planning legislation. In fact, the building code forbade all constructions on "agricultural" plots outside the city plan.

This was the case of the first "borgate" (poor neighborhoods) that were mushrooming in Rome's fringe, the "afthereta" (illegal houses) of Athens, and other illegal settlements within a distance of 20 km from the city's boundaries (as in Barcelona and Lisbon). Even if spontaneous urban development is by definition unpredictable, the pattern with which these suburbs were spreading on the territory appears to have been highly dependent on infrastructure. The first settlements were located as close as possible to roads, electricity posts, and water pipes, in order to benefit (illegally) from them. The first squatters (buildings occupied by people living in them without the legal right to do so) followed infrastructure, but surprisingly later attracted it. Governments often "gave up" to the popular pressure for infrastructure expansion in areas of illegal settlements that were mostly integrated in piecemeal ex-post legislations of city plans.

2.6 Conclusions

Having understood that urban expansion requires an *ad hoc* analysis with regards to the region of interest, the present study was dedicated to the review of urban culture and organizational model of urban development in the Mediterranean region (Zambon et al., 2018), considering urban, economic, and management variables, as well as socio-economic relationships. Urbanization is considered as the expression of cultural and intellectual characteristics of human society (Salvati et al., 2017). But an overall picture of the urban culture in the area has been particularly difficult to construct, as it has experienced many conflicts and intense exchanges from three different continents (Asia, Europe, and Africa). The conclusion is that the Mediterranean region may result as heterogeneous in some respects (i.e. spatial forms), but quite similar in others (social dynamics and behaviors). Rejecting a model of the Mediterranean city does not imply that there are not many common characteristics that urban centers of this region share.

The economic system is a crucial element in shaping the urban asset of a given area. For this reason, the analysis of the Mediterranean urban geographies starts by considering the dominant economic system of the region during the last century. The arising capitalism is seen as a contributor to urban dynamics of the region. Nevertheless, this contribution deliberately avoids going in depth with this analysis as capitalism is common to most of the Western world, and therefore does not represent a particular characteristic of the Mediterranean region. Instead, the chapter focuses on the particular social dynamics and behaviors of the Mediterranean societies, which have played a crucial role in the urban development of the region.

Urban histories of the Mediterranean feature a glorious past and a period of late industrialization combined with a demographic boom and immigration waves in the 20th century (Clemente et al., 2018; Ciommi et al., 2018; Salvia et al., 2018; Zambon et al., 2018; Cecchini et al., 2019a, 2019b; Zambon et al., 2019a, 2019b). During this period, a higher desirability for the city center and the lack of an organized industrial spatial asset (in contrast with the defined industrial zones of Anglo-American societies) has brought to the concentration of high and middle classes in core cities (Colantoni et al., 2015; Salvati and Colantoni, 2015; Carlucci et al., 2019). Nevertheless, social segregation did not have the success that it experienced in northern and central European cities (Rontos et al., 2016). Through the multistorey apartment buildings of the center, the Mediterranean cities have witnessed a social phenomenon called vertical differentiation that is, the separation of social classes on a vertical rather than horizontal axis (Brunori et al., 2018; Zambon et al., 2017, 2019a; Salvati et al., 2019a).

In the areas surrounding the center, popular land control has been the dominant trend. Due to the permissive building code and the attitude of dispensation, peripheral zones have transformed through spontaneous and illegal construction activities. This phenomenon has been observed for Italian, Portuguese, Greek, Spanish, and Turkish cities. But spatial distribution of social groups is not enough for a complete analysis about the development lines of the Mediterranean urbanization. The characteristics of specific social groups and their relationship with the economic asset of the city should be considered. For this reason, an analysis of the social formation and modes of production of the Mediterranean societies has been provided. Rather than focusing on the high social classes, the subordinate's classes (i.e. workers and poor people) in the Mediterranean area have been dominant in transforming the urban asset. These groups have acquired such an importance in the Mediterranean societies due to the strong presence of informal economy and poor industrialization of the region.

Looking at the social formation and modes of production contributes to understand the main spatial characteristics of the Mediterranean cities, based on the observation of three dimensions: industrialization, urbanization, and the city as a material context. With this approach, other important Mediterranean characteristics have been noticed. For instance, the relationship between urbanization and industrialization has led to the term "semiperipheral cities" (referring to the fact that the former has followed the latter, in contrast with northern Europe). The evidence emerging from our work is that the urban ideology of southern Europe, as cultural basis for urban management, should be considered in a historical and geographical perspective when exploring the spatial forms of the Mediterranean cities. In fact, understanding the urban ideology of a society is crucial for determining its expansion dynamics and to plan a clear management line for future development. Anglo-American and the Mediterranean societies were proposed as contrasting examples of a strong antiurban ideology against a prourban ideology.

References

Abrams, C. (1964). *Man's Struggle for Shelter in an Urbanizing World*. Cambridge, Mass: MIT Press.

Ailin, J. (2004). On the interactive relationship between urbanization and industrialization [J]. *Finance and Trade Research*, 3, 1–9.

Airoldi, G., Brunetti, G., Coda, V. (1992). *Lezioni di economia aziendale*. Bologna: Il Mulino.

Allum, P. (1973). *Politics and Society in Post-war Naples*. Cambridge: Cambridge University Press.

Amato, P. (1970). A comparison: population densities, land values and socio-economic class in four Latin American cities. *Land Economics*, 46, 447–451.

Balchin, P. (1996). *Housing Policy in Europe*. London and New York: Routledge.

Boje, D. Rosile, G. A. (2003). Comparison of socio-economic and other transorganizational development methods. *Journal of Organizational Change Management*.

Bonazzi, G. (2008). *Storia del pensiero organizzativo.* Milano: Franco Angeli. Vol. 136, 416–417.

Borgonovi, E., Marsilio, M., Musì, F. (2006). *Relazioni Pubblico privato.* Milano: Egea.

Braudel, F. (1953). *Civiltà e Imperi del Mediterraneo nell'età di Filippo II.*, Torino: Einaudi.

Braudel, F. (1987). *Il Mediterraneo Lo Spazio la Storia gli Uomini le Tradizioni.* Milano: Bompiani.

Brunori, E., Salvati, L., Antogiovanni, A., Biasi, R. (2018). Worrying about 'vertical landscapes': Terraced olive groves and ecosystem services in marginal land in central Italy. *Sustainability* (Switzerland), 104(2), 1164.

Carlucci, M., Chelli, F. M., Salvati, L. (2018). Toward a new cycle: Shortterm population dynamics, gentrification, and re-urbanization of Milan (Italy). *Sustainability* (Switzerland), 109(1), 3014.

Carlucci, M., Zambon, I., Colantoni, A., Salvati, L. (2019). Socioeconomic development, demographic dynamics and forest fires in Italy, 1961–2017: A time-series analysis. *Sustainability* (Switzerland), 115, 1305.

Cecchini, M., Cividino, S., Turco, R., Salvati, L. (2019a). Population age structure, complex socio-demographic systems and resilience potential: A spatio-temporal, evenness-based approach. *Sustainability* (Switzerland), 117, 2050.

Cecchini, M., Zambon, I., Salvati, L. (2019b). Housing and the city: A spatial analysis of residential building activity and the socio-demographic background in a Mediterranean city, 1990–2017. *Sustainability* (Switzerland), 112, 375.

Chelli, F., Gigliarano, C., Mattioli, E. (2009). The impact of inflation on heterogeneous groups of households: An application to Italy. *Economics Bulletin*, 29(2), 1276-1295.

Chelli, F.M., Ciommi, M., Emili, A., Gigliarano, C., Taralli, S. (2016). Assessing the Equitable and Sustainable Well-Being of the Italian Provinces. *International Journal of Uncertainty, Fuzziness and Knowlege-Based Systems*, 24, 39-62.

Chelli, F., Rosti, L. (2002). Age and gender differences in Italian workers' mobility. *International Journal of Manpower*, 23(4), 313-325.

Ciaccia, C., La Torre, A., Ferlito, F., Testani, E., Battaglia, V., Salvati, L., Roccuzzo, G. (2019). Agroecological practices and agrobiodiversity: A case study on organic orange in southern Italy. *Agronomy*, 92, 85.

Ciommi, M., Gigliarano, C., Emili, A., Taralli, S., Chelli, F.M. (2017). A new class of composite indicators for measuring well-being at the local level: An application to the Equitable and Sustainable Well-being (BES) of the Italian Provinces. *Ecological Indicators*, 76, 281-296.

Ciommi, M., Chelli, F. M., Carlucci, M., Salvati, L. (2018). Urban growth and demographic dynamics in southern Europe: Toward a new statistical approach to regional science. *Sustainability* (Switzerland), 108(1), 2765.

Ciommi, M., Chelli, F.M., Salvati, L. (2019). Integrating parametric and non-parametric multivariate analysis of urban growth and commuting patterns in a European metropolitan area. *Quality and Quantity,* 53(2), 957-979.

Clemente, M., Zambon, I., Konaxis, I., Salvati, L. (2018). Urban growth, economic structures and demographic dynamics: Exploring the spatial mismatch between planned and actual land-use in a Mediterranean city. *International Planning Studies*, 234(1), 376–390.

Cocklin, C., Mautner, N., Dibden, J. (2007). Public policy, private landholders: Perspectives on policy mechanisms for sustainable land management. *Journal of Environmental Management*, 85(4), 986–998.

Colantoni, A., Mavrakis, A., Sorgi, T., Salvati, L. (2015). Towards a 'polycentric' landscape? Reconnecting fragments into an integrated network of coastal forests in Rome. *Rendiconti Lincei*, 26(3), 615–624.

Colantoni, A., Zambon, I., Gras, M., Mosconi, E. M., Stefanoni, A., Salvati, L. (2018). Clustering or scattering? The spatial distribution of cropland in a metropolitan region, 1960-2010. *Sustainability* (Switzerland), 107(1), 2584.

Combes, P. P., Overman H. G. (2004). The Spatial Distribution of Economic Activities in the European Union. *Handbook of Regional and Urban Economics* 4: 2845–2909.

Corona, P., Cutini, A., Chiavetta, U., Paoletti, E. (2016). Forest-food nexus: A topical opportunity for human well-being and silviculture. *Annals of Silvicultural Research*, 40(1), 1–10.

Costa, F. J., Noble, A. G., Pendeleton, G. (1991). Evolving planning systems in Madrid, Rome, and Athens. *Geojournal*, 24(3), 293–303.

Couch, C., Leontidou, L., Petschel-Held G. (2007). *Urban Sprawl in Europe Landscapes, Land-Use Change and Policy*. Oxford, UK: Blackwell Publishing.

De Marco, A., Proietti, C., Anav, A., Ciancarella, L., D'Elia, I., Fares, S., Fornasier, M. F., Fusaro, L., Gualtieri, M., Manes, F., Marchetto, A., Mircea, M., Paoletti, E., Piersanti, A., Rogora, M., Salvati, L., Salvatori, E., Screpanti, A., Vialetto, G., Vitale, M., Leonardi, C. (2019). Impacts of air pollution on human and ecosystem health, and implications for the

National Emission Ceilings Directive: Insights from Italy. *Environment International*, 2, 320–333.

Dymski, G. A. (2005). Financial globalization, social exclusion and financial crisis. *International Journal of Applied Economics*, 19(4), 439–457.

Ferras, R. (1977). *Barcelone: Croissance d'une Metropole*. Paris: Anthropos. Fotopoulos, T. (1992). Economic restructuring and the debt problem: The Greek case. *International Journal of Applied Economics*, 6(1), 38–64.

Francaviglia, R., Di Bene, C., Farina, R., Salvati, L., Vicente-Vicente, J. L. (2019). Assessing "4 per 1000" soil organic carbon storage rates under Mediterranean climate: A comprehensive data analysis. *Mitigation and Adaptation Strategies for Global Change*, 24(5), 795–818.

Fratini, F. (2000). *Roma arcipelago di isole urbane*. Rome: Gangemi.

Fried, R. C. (1973). *Planning the Eternal City: Roman Politics and Planning since World War II*. London: Yale University Press.

Gigliarano, C., Chelli, F.M. (2016). Measuring inter-temporal intragenerational mobility: an application to the Italian labour market. *Quality and Quantity*, 50(1), 89-102.

Hall, P. (1997). The future of metropolis and its form. *Regional Studies*, 31(3), 211–220. Handlin, O. (1963). *The Historian and the City*. Cambridge: Cambridge University Press.

Ioannidis, C., Psaltis, C., Potsiou, C. (2009). Towards a strategy for control of suburban informal buildings through automatic change detection. *Computers, Environment and Urban Systems*, 33, 64–74.

Kazemzadeh-Zow, A., Zanganeh Shahraki, S., Salvati, L., Samani, N. N. (2017). A spatial zoning approach to calibrate and validate urban growth models. *International Journal of Geographical Information Science*, 31(4), 763–782.

King, R., Proudfoot, L., Smith B. (1997). *The Mediterranean. Environment and Society*. London: Arnold.

Krumholz, N. (1992). Roman impressions: Contemporary city planning and housing in Rome. *Landscape and Urban Planning*, 22(2–4), 107–114.

Legras, S. (2015). Correlated environmental impacts of wastewater management in a spatial context. *Regional Science and Urban Economics*, 52, 83–92.

Leontidou, L. (1990). *The Mediterranean City in Transition Social Change and Urban Development*. New York: Cambridge University Press.

Leontidou, L. (1996). Mediterranean cities: Divergent trends in a united Europe. In Blacksell et al. *The European Challenge: Geography and*

Development in the European Community, Oxford: Oxford University Press, 127–148.

Lynch, L., Geoghegan, J. (2011). The economics of land use change: Advancing the frontiers. *Agricultural and Resource Economics Review*, 40(3), 3–8.

Maliene, V., J. Howe, J., Malys N. (2008). Sustainable communities: Affordable housing and socio-economic relations. *Local Economy*, 23(4): 267–276.

Martinotti, G. (1993). *Metropoli*. Bologna: Il Mulino.

Monarca, D., Cecchini, M., Guerrieri, M., Colantoni, A. (2009). Conventional and alternative use of biomasses derived by hazelnut cultivation and processing. *Acta Horticulturae*, 845, 627–634.

Morelli, V. G., Rontos, K., Salvati, L. (2014). Between suburbanisation and re-urbanisation: Revisiting the urban life cycle in a Mediterranean compact city. *Urban Research and Practice*, 7(1), 74–88.

Moretti, V., Salvati, L., Cecchini, M., Zambon, I. A. (2019). Longterm analysis of demographic processes, socioeconomic 'modernization' and forest expansion in a European country. *Sustainability* (Switzerland), 112, 388.

Newman, P., Thornely A. (1996). *Urban Planning in Europe*. New York: Routledge.

Oliveira, E. (2015). Place branding in strategic spatial planning. *Journal of Place Management and Development*, 8(1), 23–50.

Proietti, C., Anav, A., Vitale, M., Fares, S., Fornasier, F., Screpanti, A., Salvati, L., Paoletti, E., Sicard, P., De Marco, A. (2019). A new wetness index to evaluate the soil water availability influence on gross primary production of European forests. *Climate*, 73, 42.

Robson, B. T. (1969). *Urban Analysis*, Cambridge: Cambridge University Press.

Rontos, K., Grigoriadis, E., Sateriano, A., Syrmali, M., Vavouras, I., Salvati, L. (2016). Lost in protest, found in segregation: Divided cities in the light of the 2015 "Oχι" referendum in Greece. *City, Culture and Society*, 7(3), 139-148.

Rosti, L., Chelli, F. (2009). Self-employment among Italian female graduates. *Education and Training*, 51(7), 526-540.

Rosti, L., Chelli, F. (2012). Higher education in non-standard wage contracts. *Education and Training*, 54(2-3), 142-151.

Salvati L. (2018). The 'niche' city: A multifactor spatial approach to identify local-scale dimensions of urban complexity. *Ecological Indicators*, 94(3), 62–73.

Salvati, L. (2019). Farmers and the city: Urban sprawl, socio-demographic polarization and land fragmentation in a Mediterranean region, 1961–2009. *City, Culture and Society*, 18, 100284.

Salvati, L., Carlucci, M., Serra, P., Zambon, I. (2019a). Demographic transitions and socioeconomic development in Italy, 1862-2009: A brief overview. *Sustainability* (Switzerland), 111, 242.

Salvati, L., Ciommi, M. T., Serra, P., Chelli, F. M. (2019b). Exploring the spatial structure of housing prices under economic expansion and stagnation: The role of socio-demographic factors in metropolitan Rome, Italy. *Land Use Policy*, 81, 143–152,

Salvati, L., Colantoni, A. (2015). Land use dynamics and soil quality in agro-forest systems: A country-scale assessment in Italy. *Journal of Environmental Planning and Management*, 58(1), 175–188.

Salvati, L., Sateriano, A., Grigoriadis, E. (2016). Crisis and the city: Profiling urban growth under economic expansion and stagnation. *Letters in Spatial and Resource Sciences*, 9(3), 329–342.

Salvati, L., Guandalini, A., Carlucci, M., Chelli, F.M. (2017). An empirical assessment of human development through remote sensing: Evidences from Italy. *Ecological Indicators*, 78, 167-172.

Salvati, L., Tombolini, I., Ippolito, A., Carlucci, M. (2018). Land quality and the city: Monitoring urban growth and land take in 76 southern European metropolitan areas. *Environment and Planning B: Urban Analytics and City Science*, 454(3), 691–712.

Salvati, L., Zambon, I., Pignatti, G., Colantoni, A., Cividino, S., Perini, L., Pontuale, G., Cecchini, M. (2019c). A time-series analysis of climate variability in urban and agricultural sites (Rome, Italy). *Agriculture* (Switzerland), 95, 103.

Salvati, L., Zitti, M., Sateriano, A. (2013). Changes in city vertical profile as an indicator of sprawl: Evidence from a Mediterranean urban region. *Habitat International*, 38, 119–125.

Salvia, R., Serra, P., Zambon, I., Cecchini, M., Salvati, L. (2018). In-between sprawl and neo-rurality: Sparse settlements and the evolution of socio-demographic local context in a Mediterranean region. *Sustainability* (Switzerland), 1010(1), 36–70.

Savall, H. (2003). An updated presentation of the socio-economic management model. *Journal of Organizational Change Management*, 16(1), 33–48.

Seronde Babonaux, A. (1983). *Roma. Dalla città alla metropoli*. Rome: Editori Riuniti.

Simon, H. A. (1947). *Administrative Behavior: A Study of Decision-making Processes in Administrative Organization*. New York: Macmillan.

Spina, M. C. (2004). *Le città del Mediterraneo*. Perugia: La Biblioteca del Mediterraneo.

Wallerstein, I. (1979). *The Capitalist World Economy*. Cambridge: Cambridge University Press.

White, P. (1984). *The West European City: a Social Geography*. London: Longman.

Williamson, O. E. (1993). Opportunism and its critics. In *Managerial and Decision Economics*, 14(2), 97–107.

Wynn, M. (1984). *Planning and Urban Growth in Southern Europe*. London: Mansell.

Zaccaria Ruggiu, A. (1995). *Spazio privato e spazio pubblico nella città romana* (Vol. 210, No. 1). Persée-Portail des revues scientifiques en SHS.

Zambon, I., Colantoni, A., Carlucci, M., Morrow, N., Sateriano, A., Salvati, L. (2017). Land quality, sustainable development and environmental degradation in agricultural districts: A computational approach based on entropy indexes. *Environmental Impact Assessment Review*, 64, 37–46.

Zambon, I., Benedetti, A., Ferrara, C., Salvati, L. (2018a). Soil matters? A multivariate analysis of socioeconomic constraints to urban expansion in Mediterranean Europe. *Ecological Economics*, 146, 173–183.

Zambon, I., Cerdà, A., Cividino, S., Salvati, L. (2019a). The (evolving) vineyard's age structure in the valencian community, Spain: A new demographicapproach for rural development and landscape analysis. *Agriculture* (Switzerland), 93, 59.

Zambon, I., Colantoni, A., Salvati, L. (2019b). Horizontal vs vertical growth: Understanding latent patterns of urban expansion in large metropolitan regions. *Science of the Total Environment*, 654(1), 778–785.

Zambon, I., Rontos, K., Serra, P., Colantoni, A., Salvati, L. (2018b). Population dynamics in southern Europe: A local-scale analysis, 1961–2011. *Sustainability* (Switzerland), 11(1), 1–12.

Zambon, I., Serra, P., Salvati, L. (2019c). The (evolving) urban footprint under sequential building cycles and changing socio-demographic contexts. *Environmental Impact Assessment Review*, 75, 27–36.

Zangrandi, A. (2019). *Aziende pubbliche*. Milano: Egea. Zanker, P. (2013). *La città romana*. Bari: Laterza.

3

Socio-Spatial Disparities and Economic Recession in the Divided Metropolis

Luca Salvati and Sabato Vinci

Urban sprawl is a socioeconomic process reflecting different political, economic, and cultural contexts. Sprawl is identifiable in suburban areas where low density residential settlements replace the traditional agricultural and forest mosaic leading to a mixed landscape, with detached family houses and where population is highly dependent on private transport (Burchell et al., 1998; Ewing et al., 2002; Tsai, 2005; Torrens, 2008; Salvati et al., 2019; Zambon et al., 2019; Sallustio et al., 2018; Carlucci et al., 2019; Moretti et al., 2019). The main causes of sprawl can be envisaged in (i) a complex system of interacting agents at the base of urban expansion (Salvati and Gargiulo Morelli, 2014); (ii) lack of efficient planning systems (Gibelli and Salzano, 2006); and (iii) a misuse of land determined by policies regulating cities' growth and the economic development of peri-urban regions (Giannakourou, 2005; Serra et al., 2018).

Sprawl can be thus considered an intriguing spatial model, involving social, economic, and environmental issues and leading to new development policies and management patterns (Burchell et al., 1998; Brueckner, 2000; Galster et al., 2001; Frenkel and Ashkenazi, 2008; Orenstein et al., 2013). Based on a large number of interacting factors, it is difficult to understand how processes of urban dispersion are structured (Kazepov, 2005; Couch et al., 2007; Cassier and Kesteloot, 2012), making it difficult to realize appropriate policies of urban containment and land-use management (Bruegmann, 2005; Hall and Pain, 2006; Angel et al., 2011; Masini et al., 2019; Bertini et al., 2019; Marchi et al., 2018; Fabbio et al., 2018). From these premises, sprawl appears to be a key issue for contemporary cities, possibly due to the overpowering degree of *laissez-faire* (Costa et al., 1991).

Southern Europe is an interesting case for studying sprawl impact on urban development (Zambon et al., 2018). The spread of low-density, scattered settlements from inner cities to suburbs is a traditional phenomenon observed since the beginning of the 20th century in North American cities (e.g. Duany et al., 2000; Bruegmann, 2005). By contrast, sprawl is a relatively new process especially in Southern Europe (European Environment Agency, 2006; Kasanko et al., 2006; Couch et al., 2007). From the 1970s, urban dispersion advanced rapidly in the Mediterranean cities, with urbanization rates growing much faster than the population. This trend was observed in Barcelona (Catalàn et al., 2008), Marseilles and the nearby Rhone valley (Pinson and Thomann, 2001), Rome (Munafò et al., 2010), and Athens (Salvati et al., 2013a), among others. The diffusion of sparse settlements driven by population de-concentration since the 1980s has created a mixed landscape around the main cities and is considered one of the most relevant socioeconomic challenges in the Mediterranean region (Alphan, 2003; Catalàn et al., 2008; Terzi and Bolen, 2009). Projections evidence the growing pressure for urbanization that Mediterranean Europe will experience in the near future (Munafò et al., 2013).

The key words emerging from this new socioeconomic context are diversification, entropy, and isolation (Burgel, 2004; Di Feliciantonio et al., 2018; Carlucci et al., 2019). Contemporary cities in Southern Europe take on a polarized spatial structure emphasizing socioeconomic disparities (Delladetsima, 2006; Leontidou et al., 2007; Maloutas, 2007; Zambon et al., 2017; Salvati 2019). The economic space became fragmented assuming a spatial organization hardly classifiable as "polycentric" (Maloutas, 2007; Colantoni et al., 2015; De Rosa and Salvati, 2016; Salvati, 2014) and devoid of services and infrastructural networks (Leontidou, 1990; Krumholtz, 1992; Giannakourou, 2005). In this work, the cases of Barcelona, Athens, and Rome are considered. These cities share similar territorial features, showing a unique structure and socioeconomic configuration. Sprawl has adapted to their contexts in different ways depending on historical, cultural, and political issues. Exhibiting a (regional or national) capital role, these cities have hosted competitive events of global importance with the Olympic Games. A narrative comparing the recent processes of urban growth reveals how sprawl has occurred and the intimate connections between compact urbanization, land use, and economic structure (Ciommi et al., 2018, 2019; Salvati et al., 2018). Integrating narrative and quantitative approaches, elements to "reset" the scene of "Southern" typologies of urban sprawl, revisiting the main socioeconomic drivers of change in contemporary cities, are provided.

3.1 Comparative Analysis of Local Contexts in the Mediterranean

The lack of unifying datasets with relevant information on a local scale complicates the comparison between metropolitan areas when identifying different sprawl types. Together with the difficulty in analyzing sprawl patterns and processes, there is a lack of basic information required to perform this type of analysis. A comprehensive picture of the recent urban development in Barcelona, Rome, and Athens is provided based on a set of indicators made available from official statistics with the aim to assess a variety of socio-economic issues related with sprawl. We used local administrative boundaries as the elementary spatial unit of analysis.

Data on the total surface of the three metropolitan areas and population density are provided in Table 3.1 using information derived from the Urban Atlas initiative of the European Environment Agency and population census. Land-use maps were also elaborated using a nomenclature based on three classes: built-up areas, cropland, and non-forest natural areas and forests (Figure 3.1). The three cities show different morphologies: compact and dense settlements in Athens, scattered and discontinuous settlements in Rome, and a more aggregated and spatially balanced urban fabric in Barcelona. Maps indicate that Athens is a relatively dense city, Rome is more chaotic, and Barcelona shows an intermediate pattern.

A morphological analysis of the three metropolitan regions was carried out considering the level of soil sealing (Figure 3.2). Soil sealing is one of the major challenges that Europe faces today because it produces negative impacts on local environments, reflecting the "footprint" of each city (Munafò et al., 2013). Results bring evidence to the compactness of Athens' morphology (Table 3.2), the spatial discontinuity of Rome's settlements, and the polarization in high and low-density settlements when approaching Barcelona's expansion (Carlucci et al., 2017).

Table 3.1 Basic characteristics of the three cities

City	Surface area (km^2)	Population	Population density (inhab/km^2)
Barcelona	3242	4,394,412	1355
Rome	5352	3,700,424	691
Athens	3025	3,724,393	1231

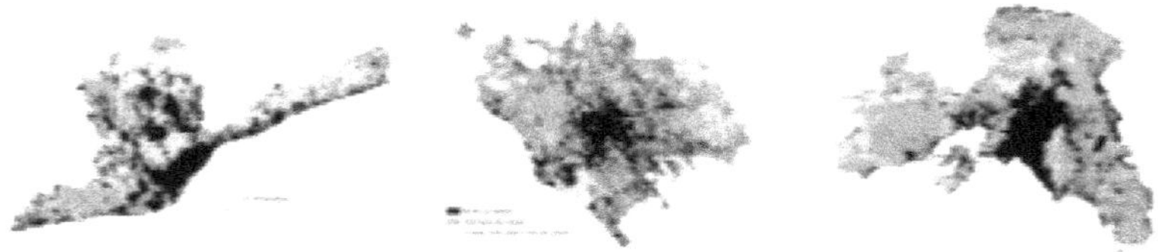

Figure 3.1 Land-use maps in Barcelona (left), Rome (middle), and Athens (right).

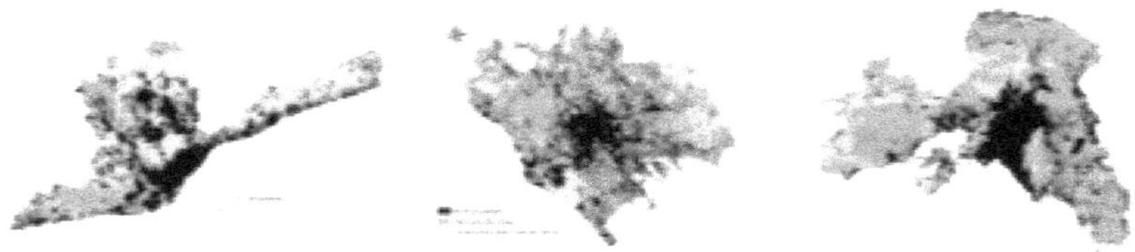

Figure 3.2 Percent degree of soil sealing in the three cities.

Table 3.2 Land cover by soil sealing class (%) in the three cities

Class	Barcelona	Rome	Athens
0	72.0	76.5	75.4
1-20	9.9	10.9	8.1
21-40	5.1	4.9	4.5
41-60	3.6	2.8	3.5
>60	9.4	4.9	8.5

3.1.1 Population Growth and Urban Expansion

Three indicators were considered to assess local-scale population growth in the short term (2001-2011) and in the long term (1981-2011) as well as population density as a result of urban concentration (or dispersion). Indicators evaluate the relocation of population in the outer fringe as a consequence of sprawl dynamics (Figure 3.3). Barcelona experienced spatially diffused population growth in the last decade and a demographic stability with scattered expansion in the last three decades. Population growth in Rome was diffused along the radial axes opposed to the inner city, reproducing a fragmented landscape at distances progressively further away from central districts (Salvati et al., 2013b). In the last 30 years, municipalities with the largest increase in resident population were situated in the outer ring of Rome, particularly along coastal areas. The central city showed a moderate decrease in resident population, more pronounced in the last decade. Population growth in Athens was relatively stable over time, although with a moderate decrease in the central city, reflecting the polarization in shrinking industrial

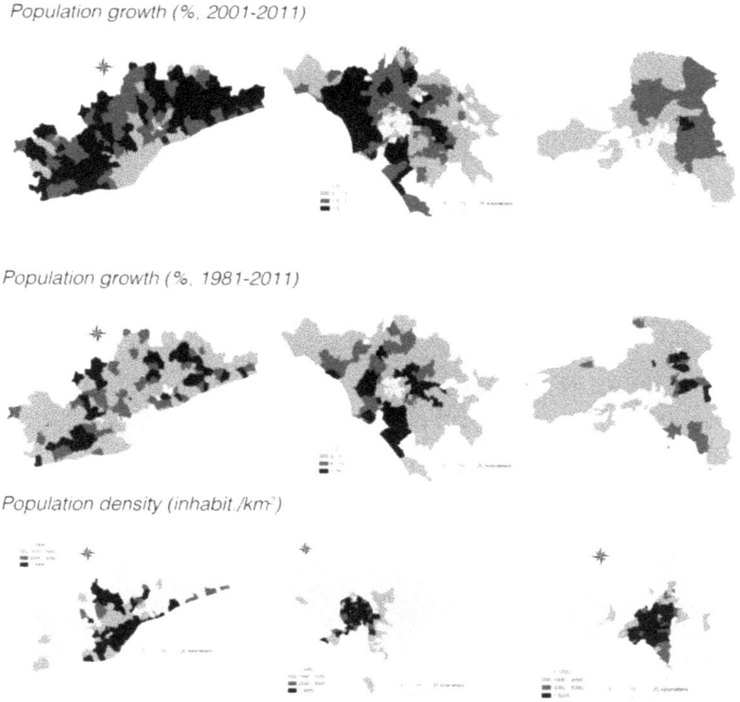

Figure 3.3 Recent population dynamics in the three cities.

areas (Piraeus and surrounding districts) and sprawling eastern suburbs close to the international airport. Population density discriminates

Barcelona and Athens (more compact) from Rome (discontinuous, medium-density settlements). Four indicators were calculated and mapped with the aim at describing the dominant social context in the three cities (Figure 3.4): (i) the percent share of one-component families in the total number of households; (ii) an index of population aging; (iii) the percent share of foreigners in total resident population; and (iv) the percent share of graduates in total resident population.

The average number of components per family is rather similar in the three cities. Barcelona has values ranging from 1.7 to 2.8, with prevalence of medium to small households, with an average of 2.4 components over the metropolitan area. Athens shows moderately higher values ranging between 2.3 and 4.0, with 3.2 components per family on average. Rome ranks in the middle with average household size amounting to 2.9. In all the cities, large households tend to settle in suburban areas. By contrast, mononuclear households settled prevalently in urban areas. Elderly index completes the picture illustrating a peri-urban distribution of elderly population around the central city of Barcelona, while inland, rural municipalities showed high values of the index, together with some urban districts, in Rome. In Athens, the highest elderly index was observed in the urban area and, more scattered, in some coastal and inland peri-urban municipalities. The spatial distribution of the index well reflects the dominant demographic patterns in the three cities, possibly influencing the overall process of sprawl. The proportion of foreign resident population was also discriminant among cities: Athens hosts the largest foreign population when compared with Barcelona and Rome. By contrast, the spatial distribution of graduates (tertiary education) follows a typical spatial gradient in all cities, being higher in central cities and lower in peri-urban areas (Serra et al., 2018). However, differences were observed in Barcelona and Athens with a more polarized distribution of graduates in the former city compared with the latter.

The economic structure of the three cities was studied considering four key elements: (i) unemployment rate, (ii) average declared income, (iii) share of entrepreneurs in total resident population, and (iv) percent share of workers in credit, insurance, and business services in total workforce (Figure 3.5). The unemployment rate is higher in Barcelona (58% on average); in Rome, several municipalities had a low or medium-low employment rate, while in Athens high employment rates were observed in Eastern Attica. Unemployment rate was lower in Barcelona and much higher in Rome while in Attica an east-west divide was observed with unemployed population concentrated in western industrial districts.

Together with the unemployment rate, average per capita declared income is a key indicator when assessing the socioeconomic context, influencing sprawl in the Mediterranean cities. In Barcelona, the municipalities with the highest income (> 25,000 euros per-capita) are situated close to Barcelona or along the coastline. The municipality of Barcelona has intermediate values of per-capita income. In Rome, the highest incomes are concentrated around the consolidated city expanding along the sea coast. In Athens, the highest incomes were found in the eastern part of the region reflecting socioeconomic disparities between western and eastern districts. Barcelona showed a relatively high density of entrepreneurs settled in urban areas and in some suburban municipalities. In Rome, entrepreneurs concentrated in suburban settlements especially along the coastline. The maximum value of the indicator was 0.5 in Rome, declining to 0.2 and

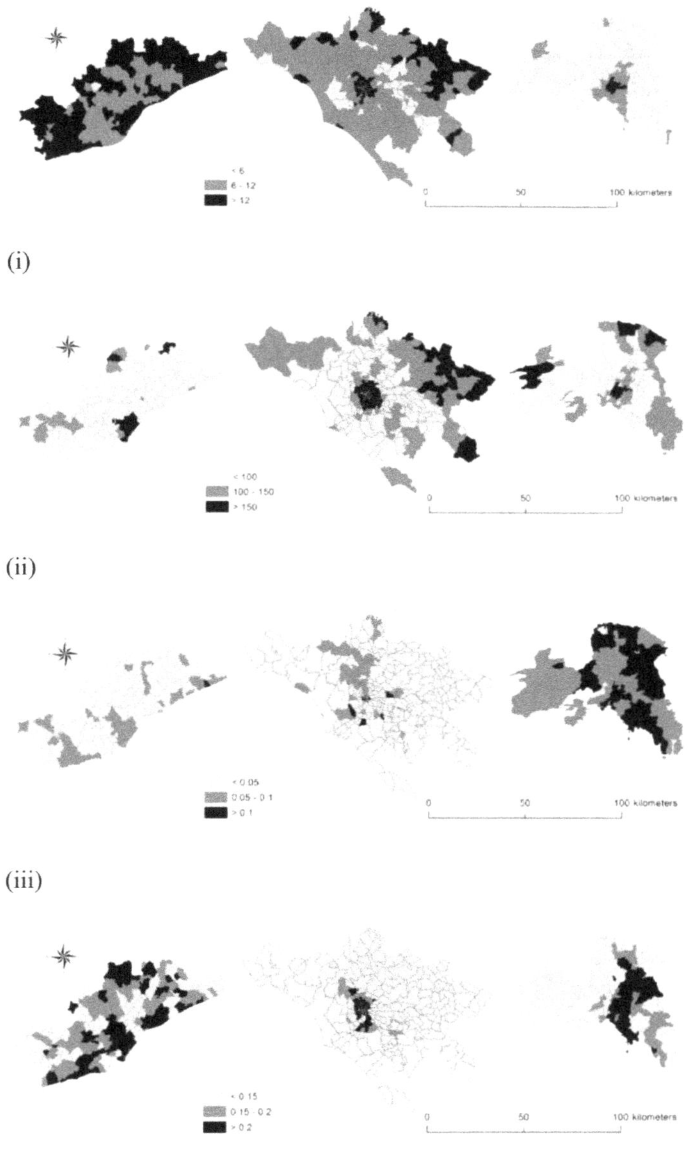

Figure 3.4 Social indicators in the three cities: (i) percent share of one-component families in the total number of households; (ii) index of population aging; (iii) percent share of foreigners in total resident population; and (iv) percent share of graduates in total resident population.

0.3, respectively in Athens and Barcelona. Entrepreneurs concentrate in the eastern district of Athens outlining moderate class segregation (Rontos et al., 2016). A similar spatial distribution in the three cities was observed for the share of workers in bank and insurance services in total workforce. In the case of Barcelona, the highest values of the indicator were found in coastal municipalities, possibly due to tourism, real estate, and second-home expansion (Cuadrado-Ciuraneta et al., 2017). A higher dispersion was found in Rome, with a moderate increase in the central city. In Athens, the eastern districts attract the highest proportion of workers in financial services.

3.2 Dispersed Urbanization in the Mediterranean Area as Public Governance Trend

3.2.1 Barcelona: Toward a Scattered-polycentric Region?

Urban sprawl in Spain, and particularly Catalonia, has had multifaceted patterns and dynamics. The largest Spanish cities share factors such as the concentration of upper functions and dense job markets, high rate of commuting, uneven expansion of the urban fabric, and a progressive functional specialization of the different spaces within metropolitan areas. Urban expansion is taking place in different ways in Spanish regions due to cultural diversity and territorial heterogeneity. The Barcelona metropolitan region is composed of the central city (partitioned into 10 urban districts), and 163 peri-urban municipalities, each claiming a distinctive identity and with different social, political, and economic contexts. In addition, the topography has played a major role in the evolution of metropolitan form, because of its homogeneity, being dominated by two mountain ranges (the littoral range, with elevation above 700 m, and the pre-littoral range, with elevation above 1700 m) and two flat areas. The spatial distribution of population in the metropolitan area of Barcelona is different from what was observed for other European metropolitan regions. After two decades (1960–1980) of high immigration from rural areas of Southern Spain, population of Barcelona has grown from 1.56 to 1.75 million, while in 11 towns around the central city the population has grown more rapidly from 0.57 to 1.41 million inhabitants. The resulting suburbanization combines high-density areas and industrial settlements, an irregular urban structure and population concentration in the inner city. Since the 1950s, the Barcelona's metropolitan region has experienced an accelerated process of land consumption (Salvati et al., 2018).

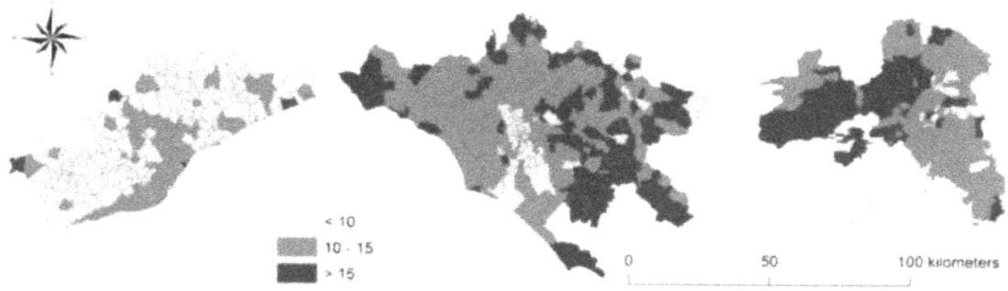

(i)

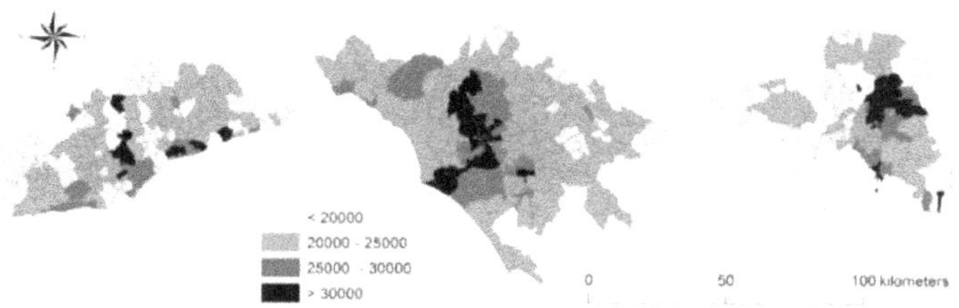

(ii)

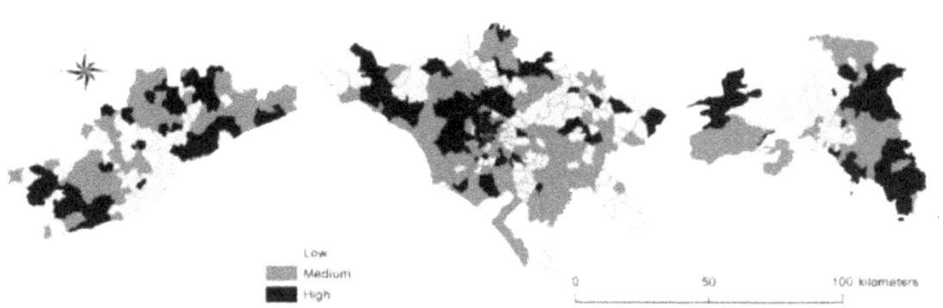

(iii)

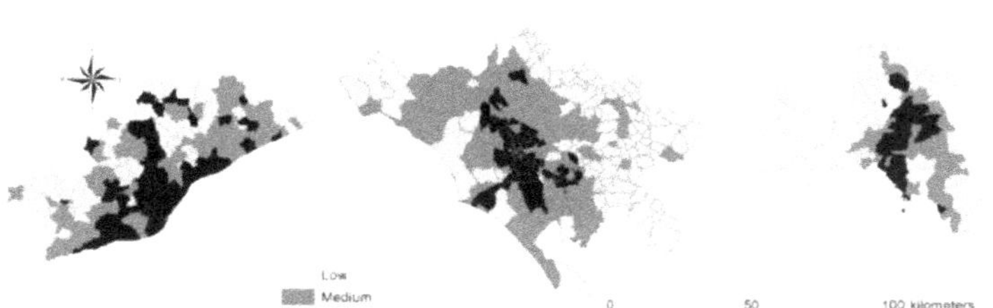

(iv)

Figure 3.5 Economic indicators in the three Mediterranean cities: (i) unemployment rate; (ii) average declared income; (iii) share of entrepreneurs in total resident population; and (iv) share of workers in credit, insurance, and business services in total workforce.

Between 1950 and 1975, new neighborhoods were built-up at the edge of traditional cities to accommodate the increasing population moving from rural areas. Urban development created peripheries with serious shortcomings of open spaces, infrastructures, social services, and public transportation. At the same time, the growing congestion in the inner city was boosted by construction of second homes, often on cheap and illegal land of the outer fringes. Since 1975, a stabilization of population flows and migration was witnessed even if the residential dispersion continued (Catalan et al., 2008). Between 1975 and 1986, coinciding with a period of economic recession in Spain, urban growth was moderate. Land occupation somewhat receded, but residential decentralization was sustained through infilling former urban "islands." Many of the second homes built-up during the previous time period became main residences consolidating dispersed settlements (Cuadrado-Ciuraneta et al., 2017). Local economy began to recover in 1986, and a new wave of land occupation in the metropolitan peripheries took place during the 1990s (Betteley et al., 2000).

Metropolitan landscape is the result of a historical process adding successive layers that combine in different development patterns and schemes of metropolitan management. The general trend has been an expansion to the outside of the metropolitan region, with the most rapid growth at the margins but not in simple concentric rings. Some authors have claimed that Spanish sprawl is similar to the North American suburban model in terms of low-density housing typologies and the new residential landscapes. By contrast, other scholars state that (i) in many Spanish cities sprawl is a relatively recent phenomenon; (ii) inner districts in the Mediterranean cities have retained high population density; and (iii) due to the traditional compact nature of Southern European cities, morphology and intensity of sprawl are different from the American model evidencing a marked urban-rural divide (Carlucci et al., 2018; Ciommi et al., 2018, 2019; Salvati et al., 2017, 2019).

Muñoz (2003) stated that 40% and 70% of new houses, respectively in the metropolitan area of Barcelona and in fringe towns were detached family houses. The relocation of industrial activities was another factor contributing to the expansion of scattered settlements in turn stimulating car ownership. A number of industries have moved from the central area of Barcelona to the outer parts of the metropolitan region. An increase in the activities of retail sales was observed in fringe districts. Large shopping centers often arose in areas that offer cheap land, with easy access to transport networks, and consolidate the classic leapfrogging pattern of sprawl.

The current saturation of the inner core limits future growth of Barcelona and its conurbation. Sub-centers have an important historical background and high population density, while neighboring municipalities offer affordable housing prices and modest densities showing the highest potential for growth (Mariani et al., 2018). Land occupation, loss in agricultural and forest cover, decrease in urban densities, and the high amount of bare land expecting for further development are important signals of landscape transformation and robust indicators of the new directions taken by urbanization (Prokopová et al., 2018; Sallustio et al., 2018; Salvati et al., 2019; Zambon et al., 2019). Land-use changes in the area testify an outward expansion of the consolidated city thanks to the high level of car ownership, the relocation of industrial and retailing activities to the fringe, the development of transport infrastructure, and the conversion of second homes into primary residences (e.g. Gigliarano and Chelli, 2016; Cuadrado-Ciuraneta et al., 2017).

3.2.2 Rome: Urban Sprawl and Uncertain Socio-economic Development

Rome metropolitan area extends 5,355 km^2 of land encompassing the Functional Urban Area of Rome. The area around the city (the so-called "Agro Romano") includes the alluvial plain of the Tiber River with fertile soils traditionally used for horticulture and tree crop (Biasi et al., 2015). In a public governance point of view, the region is administered by 122 municipalities, among which Rome municipality was partitioned into 19 urban districts covering a surface area of 1285 km^2. Rome's municipality includes a heterogeneous territory, with mixed impervious and semi-natural land contrasting the compactness of the historical center, where the most important functions are concentrated. Two demographic phases have been identified in Rome along the recent growth path: a "compact expansion" covering the time interval between World War II and the late 1970s, and a "dispersed expansion" since the early 1980s. Population grew in urban areas at a higher rate than the suburban area (3.3% vs 1.8% per year) in the former period. In the latter phase population declined in the urban area (−0.2% per year) while rising in suburban areas at a rate of 1–2% per year, on average (Munafò et al., 2013).

During the compact expansion, Rome's socioeconomic structure was characterized by high unemployment rate, low activity rate, and an economic structure based on commerce, public services, and constructions (Seronde Babonaux, 1983; Costa et al., 1991; Krumholz, 1992; Insolera, 1993). Since the 1980s, a moderate de-concentration of the inner city was observed. The growth of dispersed, medium-density settlements was rapid, while economic structure and social disparities were oriented along the urban gradient (Salvati et al., 2013a; Zambon et al., 2017). In addition, during the last 20 years, investments were concentrated in the inner city, depressing the developmental potential of peripheral areas (Gonzales, 2011). The creation of a highly fragmented suburban landscape around Rome did not significantly affect the socioeconomic characteristics of fringe districts (Allegretti and Cellamare, 2008). However, land consumption in Rome was high during both "compact growth" (due to population increase), and in the most recent time interval characterized by demographic stability (Salvati et al., 2015, 2018). Salvati and Sabbi (2011) illustrated the progressive consolidation of a scattered and chaotic urban fabric dominated by discontinuous and low-density settlements at higher distances from the inner city.

Despite urban dispersion, Rome is a city with high levels of urban congestion, population concentration, and economic polarization (Seronde Babonaux, 1983). Resident population decreased only recently in core municipalities, as observed in other Mediterranean cities (Durà-Guimera, 2003; Phelps et al., 2006; Couch et al., 2007; Chorianopoulos et al., 2010; Salvati and Gargiulo Morelli, 2014). The actual structure of Rome's metropolitan area is the result of a scattered and chaotic development. Rome's Master Plan (1993–2008) identified tourism and culture as the two main sectors promoting urban development. Measures for polycentric development were introduced promoting sub-central development (Gemmiti et al., 2012; Mariani et al., 2018), strengthening some of the existing metropolitan poles, stimulating new economic activities, and relocalizing urban functions (Allegretti and Cellamare, 2008). However, this strategy was applied partially or revealed poorly effective overall (Munafò et al., 2013).

3.2.3 Athens, Sprawled by Chance?

Attica is one of the historical regions of Greece, which includes its own peninsula that juts into the Aegean Sea and where the Greek capital, Athens, is located. The region covers nearly 3800 km^2 and is administered by 60 municipalities. The municipality of Athens included 650,000 inhabitants in 2011, which added to the 3 million inhabitants of the metropolitan area at large (Greater Athens). Athens' development is a typical example of the informal growth path followed by several Mediterranean cities, which are transforming from compact to dispersed settlement models. The interest in providing a detailed analysis of the urban development of Athens in the 20th century derives from various reasons. Athens is one of the few large urban areas of Europe still manifesting a relatively intense demographic increase, followed by a drastic over-urbanization process and high rates of land consumption (Salvati et al., 2018), which is transforming the typical landscape of Attica. Athens also represents a good example for studying the impact of mega-events on urban development. At the same time, the Olympic Games, besides placing Greece under the spotlight of the world's eyes, have made the city an interesting example for discussing how new entrepreneurial city modes are influencing urbanization.

Exhibiting similar traits with Spanish and Italian cities, Athens economic structure in the postwar period was based on urbanization economies that triggered a slow process of industrialization. As a result, the city did not experience the usual deindustrialization/disurbanization wave observed in Northern Europe since the 1970s, growing in an un-designed fashion. Unplanned expansion was based on self-financed property development, with limited public expenditure for urban infrastructure (Economou et al., 2007). Urban development mainly occurred through self-built housing, sprawling onto cheap suburban land with severe infrastructure deficiencies (e.g. lack of water and sewage systems, streets, public transport, and social amenities). This led to social exclusion and vulnerability of the resident population, not to mention the environmental deterioration of the landscape.

The metropolitan region of Athens has been characterized by a marked urban–rural divide. However, throughout the 20th century, spatial polarizations have been rapidly evolved, owing to exurban development, immigration, and a weak planning system combined with a "permissive" building code. These processes have modified the spatial distribution of economic activities in Attica (Kourliouros, 1997). The high population density in the inner city (nearly 20,000 inhabitants per km^2) has led to a reorganization of the intimate structure of the metropolitan region. The increasing demand for construction sites destined to commercial, industrial, residential, and recreational activities was leading to an inevitable expansion of the urban area beyond its traditional boundaries. In the 1980s, workers and popular strata, together with rural immigrants, moved to the surroundings of Athens and Piraeus in search of affordable housing. Consequently, the density gaps between central cities and rural areas declined rapidly. The urban–rural depolarization has made possible thanks to the massive infrastructural development and the permissive urban policy. In the 1990s, middle- and upper-class exodus from inner cities became significant. The 2004 Olympic Games have played a crucial role in shaping sprawl at the regional scale. Urban competitiveness for natural and financial resources for the Olympic Games has grown more than ever, giving space to the entrepreneurial city model. Mega-events are intended as a means of self-promotion, stimulating infrastructure development, and strategies for enhancing the image of the hosting city.

When Greece was awarded with the Olympic Games, several urban and infrastructural projects were started. These were aimed at reducing peripherality and improving the functional aspects and image of the metropolis (i.e. upgrade of the underground and suburban railway, expansion of the international airport, and connection of the archaeological sites). The Games acted as a "catalyst" for the reorientation of the space policy toward the improvement of urban landscapes. However, the event has had two main consequences: the emergence of new spatial links between urban and rural areas and changes in real estate dynamics boosting sprawled urban expansion. Furthermore, these interventions have been distributed throughout Athens, determining a multinucleated regeneration program (Beriatos and Gospodini, 2004). Consequently, ribbon and leapfrog sprawl has been encouraged. Compared with Barcelona and Rome, Athens displays a mono-centric structure with the inner city maintaining its role as an economic attractor (Salvati and De Rosa, 2014), despite the presence of specific areas intended exclusively for low-density settlements (Kourliouros, 1997).

3.3 Conclusions

Our analysis shows how socioeconomic structures diverge in the three urban regions, influencing their development lines and management models. Sprawl has occurred in each of the three contexts but has adapted in different ways, following economic and social forces. The combination of various factors has allowed the development of new paths of dispersed urbanization (European Environment Agency, 2006). Contemporary cities expanding into metropolitan regions are destined to emerge with a competitive and innovative self-image (Longhi and Musolesi, 2007; Turok and Mykhnenko, 2007; Schneider and Woodcock, 2008; Fregolent and Tonin, 2013). For instance, Barcelona and Athens were promoted as places of megaevents, such as the Olympic Games (Leontidou et al., 2007). Cities have invested in their own territory in order to appear attractive and to improve the competitiveness of their metropolitan areas (Chorianopoulos et al., 2010; Pagonis, 2013).

Investments were dispersed throughout the metropolitan area, flowing out new territorial transformations (Richardson and Chang-Hee, 2004;

Bruegmann, 2005; Phelps et al., 2006; Catalàn et al., 2008). Transformations have been successful in promoting suburban lifestyles (Durà-Guimera, 2003; Muñoz, 2003; Marucci et al., 2013; Salvati et al., 2013b; Boubaker et al., 2014). The present contribution suggests that sprawl outcomes are associated with socioeconomic issues, being influenced by territorial dynamics at large.

Although Barcelona, Rome, and Athens experienced different urbanization paths in the last decades, these cities have all seen the growth of dispersed settlements, driven by some forms of planning deregulation. Territorial dynamics have shaped the economic structure of the three cities. Barcelona polycentric structure reflects the consolidation of urban sub-centers scattered around the central city (Mariani et al., 2018). The metropolitan area of Rome is more entropic and morphologically "scattered," highlighting how low-density settlements grew spontaneously up to the 1980s (Krumholz, 1992; Salvati and De Rosa, 2014). Finally, Athens maintains its role as capital city consolidating a typical monocentric form, despite the presence of specific areas destined to low-density settlements (Leontidou, 1990; Kourliouros, 1997; Couch et al., 2007). Dispersed settlements have led to economic polarizations and social homogenization (Leontidou, 1996; Pacione, 2003; Beriatos and Gospodini, 2004; Carlucci et al., 2017). Socio-spatial disparities had also a negative impact on local cohesion and sense of belonging, considered as one of the outcomes of sprawl (Gibelli and Salzano, 2006; Salvati and Colantoni, 2015; Zambon et al., 2017). Based on the results of this study and on previous works, we can only conclude that the negative consequences of sprawl are mainly related to (i) changes in the use of land destined for low density residential settlements, (ii) changes in urban lifestyles toward social homogenization and individualism; and (iii) a moderate loss of economic attractiveness of inner cities partly counterbalanced with a gaining importance of sub-centres. The recent experience of Barcelona, Rome, and Athens, and the role of local contexts shaping sprawl and socioeconomic disparities are seen as exemplificative of the diverging metropolitan dynamics observed in Southern Europe (e.g. Rosti and Chelli, 2012; Chelli et al., 2016; Ciommi et al., 2017. Integrated land management is required to consider the diversity of sprawl typologies as an element of urban complexity (Fabbio et al., 2018; Marchi et al., 2018; Bertini et al., 2019; Masini et al., 2019).

References

Allegretti G., Cellamare C. (2009). *The ambiguous renaissance of Rome*. In Porter L., Shaw K. (Eds.), *Whose Urban Renaissance?* London: Routledge.

Alphan H. (2003). Land use change and urbanization of Adana, Turkey. *Land Degradation and Development*, 14, 575–586.

Amin, A., Thrift, N. (2000). *Riflessioni sulla competitività della città*. Foedus, 1, 5–25.

Anas, A. (2013). *Modelling in Urban and Regional Economics*. Taylor & Francis.

Angel, S., Parent, J., Civco, D. L., Blei, A., Potere, D. (2011). The dimensions of global urban expansion: Estimates and projections for all countries, 2000–2050. *Progress in Planning*, 75(2), 53-107.

Beriatos, E., Gospodini, A. (2004). "Glocalising" urban landscapes: Athens and the 2004 Olympics. *Cities*, 21, 187–202.

Bertini, G., Becagli, C., Chiavetta, U., Ferretti, F., Fabbio, G., Salvati, L. (2019). Exploratory analysis of structural diversity indicators at stand level in three Italian beech sites and implications for sustainable forest management. *Journal of Forestry Research*, 301, 121–127.

Betteley, D. D, Valler, D. (2000). Integrating the economic and the social: Policy and institutional change in local economic strategy. *Local Economy*, 14(4), 295–312.

Biasi, R., Brunori, E., Smiraglia, D., Salvati, L. (2015). Linking traditional tree-crop landscapes and agro-biodiversity in Central Italy using a database of typical and traditional products: A multiple risk assessment through a data mining analysis. *Biodiversity and Conservation*, 24(12), 3009–3031.

Boubaker, K., Colantoni, A., Allegrini, E., Longo, L., Di Giacinto, S., Monarca, D., Cecchini, M. (2014). A model for musculoskeletal disorder-related fatigue in upper limb manipulation during industrial vegetables sorting. *International Journal of Industrial Ergonomics*, 44(4), 601–605.

Bruegmann, R. (2005). *Sprawl: A Compact History*. Chicago: University of Chicago Press.

Burchell, R. W., Shad, N. A., Listokin, D., Phillips, H., Downs, A., Seskin, S., Davis, J. S., Moore, T., Helton, D., Gall, M. (1998). *The Costs of Sprawl-Revisited*. Report 39. Transit Cooperative Research Program, Transportation Research Board, National Research Council. Washington, DC: National Academy Press.

Burgel, G. (2004). Athènes, de la balkanisation à la mondialisation. *Méditerranée*, 103(3-4), 59–63.

Camagni, R., Gibelli, M. C., Rigamonti, P. (2002). *I costi collettivi della città dispersa*. Firenze: Alinea.

Carlucci, M., Grigoriadis, E., Rontos, K., Salvati, L. (2017). Revisiting a hegemonic concept: Long-term 'Mediterranean urbanization' in between city re-polarization and metropolitan decline. *Applied Spatial Analysis and Policy*, 10(3), 347–362.

Carlucci, M., Chelli, F.M., Salvati, L. (2018). Toward a new cycle: Short-term population dynamics, gentrification, and re-urbanization of Milan (Italy). *Sustainability (Switzerland)*, 10(9), 3014.

Carlucci, M., Zambon, I., Colantoni, A., Salvati, L. (2019). Socioeconomic development, demographic dynamics and forest fires in Italy, 1961–2017: A time-series analysis. *Sustainability* (Switzerland), 115, 1305.

Cassiers, T., Kesteloot, C. (2012). Socio-spatial inequalities and social cohesion in European cities. *Urban Studies*, 49(9), 1909–1924.

Catalàn, B., Sauri, D., Serra, P. (2008). Urban sprawl in the Mediterranean? Patterns of growth and change in the Barcelona Metropolitan Region 1993–2000. *Landscape and Urban Planning*, 85(3–4), 174–184.

Chelli, F.M., Ciommi, M., Emili, A., Gigliarano, C., Taralli, S. (2016). Assessing the Equitable and Sustainable Well-Being of the Italian Provinces. *International Journal of Uncertainty, Fuzziness and Knowlege-Based Systems*, 24, 39-62.

Chorianopoulos, I., Pagonis, T., Koukoulas, S., Drymoniti, S. (2010). Planning, competitiveness and sprawl in the Mediterranean city: The case of Athens. *Cities*, 27, 249–259.

Ciommi, M., Gigliarano, C., Emili, A., Taralli, S., Chelli, F.M. (2017). A new class of composite indicators for measuring well-being at the local level: An application to the Equitable and Sustainable Well-being (BES) of the Italian Provinces. *Ecological Indicators,* 76, 281-296.

Ciommi, M., Chelli, F. M., Carlucci, M., Salvati, L. (2018). Urban growth and demographic dynamics in southern Europe: Toward a new statistical approach to regional science. *Sustainability* (Switzerland), 108, 27–65, 1.

Ciommi, M., Chelli, F.M., Salvati, L. (2019). Integrating parametric and non-parametric multivariate analysis of urban growth and commuting patterns in a European metropolitan area. *Quality and Quantity*, 53(2), 957-979.

Colantoni, A., Mavrakis, A., Sorgi, T., Salvati, L. (2015). Towards a 'polycentric' landscape? Reconnecting fragments into an integrated network of coastal forests in Rome. *Rendiconti Lincei*, 26(3), 615–624.

Costa, F., Noble, A. G., Pendleton, G. (1991). Evolving planning systems in Madrid, Rome, and Athens. *Geojournal*, 24, 293–303.

Couch C., Petschel-held G., Leontidou L. (2007). *Urban Sprawl in Europe: Landscapes, Land-use Change and Policy*. London: Blackwell.

Cuadrado-Ciuraneta, S., Durà-Guimerà, A., Salvati, L. (2017). Not only tourism: Unravelling suburbanization, second-home expansion and "rural" sprawl in Catalonia, Spain. *Urban Geography*, 38(1), 66–89.

De Rosa, S., Salvati, L. (2016). Beyond a 'side street story'? Naples from spontaneous centrality to entropic polycentricism, towards a 'crisis city'. *Cities*, 51, 74–83.

Delladetsima, P. M. (2006). The emerging property development pattern in Greece and its impact on spatial development. *European Urban and Regional Studies*, 13(3), 245–278.

Di Feliciantonio, C., Salvati L., Sarantakou, E., Rontos, K. (2018). Class diversification, economic growth and urban sprawl: Evidences from a pre-crisis European city. *Quality and Quantity*, 524(3), 1501–1522.

Domènech, E., Saurí, D. (2006). Urbanization and water consumption: Influencing factors in the metropolitan region of Barcelona. *Urban Studies*, 43(9), 1605–1623.

Dura-Guimera, A. (2003). Population deconcentration and social restructuring in Barcelona, a European Mediterranean city. *Cities*, 20, 387–394.

Economou, D., Petrakos, G., Psycharis, Y. (2007). *Urban policy in Greece*. In: *National Policy Responses to Urban Challenges in Europe*. L. Van den Berg, E. Braun J. Van der Meer (Eds.). Aldershot: Ashgate.

European Environment Agency (2006). *Urban sprawl in Europe – The ignored challenge.* Copenhagen: EEA Report No. 10.

European Environment Agency (2010). Mapping guide for a European urban atlas. Copenhagen: Version 1.1. EEA.

Ewing, R., Pendall, R., Chen, D. (2002). *Measuring sprawl and its impact.* Volume 1, technical report, Smart Growth America, Washington, DC.

Fabbio, G., Cantiani, P., Ferretti, F., dDi Salvatore, U., Bertini, G., Becagli, C., Chiavetta, U., Marchi, M., Salvati, L. (2018). Sustainable land management, adaptive silviculture, and new forest challenges: Evidence from a latitudinal gradient in Italy. *Sustainability* (Switzerland), 107(1), 2520.

Fregolent, L. (2011). La diversitat d'escenaris en la gestió de la ciutat de baixa densita: Experiències internacionals. In: Munoz F. (Ed.), *Estratègies per a la ciutat de baixa densitat. De la contenció a la gestió*. Barcelona: Diputació de Barcelona, 449–462.

Fregolent, L., Tonin, S. (2011). *Lo sviluppo urbano disperso e le implicazioni sulla spesa pubblica. Economia e società regionale.* Oltre il Ponte, 112, 41–60.

Frenkel, A., Ashkenazi, M. (2008). Measuring urban sprawl: How can we deal with it? *Environment and Planning B: Planning and Design*, 35, 56–79.

Gaffikin, F., Perry, D. C. (2012). The contemporary urban condition: Understanding the globalizing city as informal, contested, and anchored. *Urban Affairs Review*, 48(5), 701–730.

Garcia-López, M. A., Muniz, I. (2010). Employment decentralisation: Polycentricity or scatteration? The case of Barcelona. *Urban Studies*, 47(14), 3035–3056.

Gemmiti, R., Salvati, L., Ciccarelli, S. (2012). Global city or ordinary city? Rome as a case study. *International Journal Latest Trends in Finance and Economic Science*, 2(2), 91–98.

Giannakourou, G. (2005). Transforming spatial planning policy in Mediterranean countries: Europeanization and domestic change. *European Planning Studies*, 13, 319–331.

Gibbs, D. (1997). Urban sustainability and economic development in the United Kingdom: Exploring the contradictions. *Cities*, 14(4), 203–208.

Gibelli, M.C., Salzano, E. (2006). *No Sprawl*. Firenze: Alinea.

Gigliarano, C., Chelli, F.M. (2016). Measuring inter-temporal intragenerational mobility: an application to the Italian labour market. *Quality and Quantity*, 50(1), 89-102.

Glaster, G., Hanson, R., Ratcliffe, M. R., Wolman, H., Coleman, S., Freihage, J. (2001). Wrestling sprawl to the ground: Defining and measuring an elusive concept. *Housing Policy Debate*, 12(4), 681–717.

Gonzales, S. (2011). The north/south divide in Italy and England: Discursive construction of regional inequality. *European Urban and Regional Studies*, 18(1), 62–76.

Gordon P., Richardson, H. W. (2000). Are compact cities a desirable planning goal? *Journal of the American Planning Association*, 63(1), 89–106.

Hall, P., Pain, K. (2006). *The Polycentric Metropolis. Learning from Megacity Regions in Europe*. London, UK: Earthscan.

Insolera, I. (1993). *Roma moderna. Un secolo di urbanistica romana 1870–1970*. Turin: Einaudi.

Jensen, J. R., Cowen D. C. (1999). Remote sensing of urban/suburban infrastructure and socioeconomic attributes. *Photogrammetric Engineering and Remote Sensing*, 65, 611–622.

Kazepov, Y. (2005). *Cities of Europe: Changing Contexts, Local Arrangements, and the Challenge to Urban Cohesion*. Oxford, UK: Blackwell.

King, R., Proudfoot, L., Smith, B. (1997). *The Mediterranean. Environment and Society*. London: Arnold.

Kourliouros E. (1997). Planning industrial location in greater Athens: The interaction between deindustrialization and anti-industrialism during the 1980s. *European Planning Studies*, 5(4), 435–460.

Kresl, P. K. (2007). *Planning Cities for the Future. The Successes and Failures of Urban Economic Strategies in Europe*. Cheltenham: Edward Elgar.

Krumholz, N. (1992). Roman impressions: Contemporary city planning and housing in Rome. *Landscape and Urban Planning*, 22(2–4), 107–114. Lambooy, J. G. (1998). Polynucleation and economic development: The Randstad. *European Planning Studies*, 6(4), 457–467.

Leontidou, L. (1990). *The Mediterranean City in Transition*. Cambridge: Cambridge University Press.

Leontidou, L. (1996) Alternatives to modernism in (Southern) urban theory: Exploring in-between spaces. *International Journal of Urban and Regional Research*, 20(2), 180–197.

Longhi, C., Musolesi, A. (2007). European cities in the process of economic integration: Towards structural convergence. *Annals of Regional Science*, 41, 333–351.

Maloutas, T. (2007). Socioeconomic classification models and contextual difference: The "European socioeconomic classes" (ESeC) from a south European angle. *South European Society and Politics*, 12(4), 443–460.

Marchi, M., Ferrara, C., Biasi, R., Salvia, R., Salvati, L. (2018). Agro-forest management and soil degradation in Mediterranean environments: Towards a strategy for sustainable land use in vineyard and olive cropland. *Sustainability* (Switzerland), 107(2), 25–65.

Mariani, F., Zambon, I., Salvati, L. (2018). Population matters: Identifying metropolitan sub-centers from diachronic density-distance curves, 1960–2010. *Sustainability* (Switzerland), 1012, 46–53.

Marucci, A., Zambon, I., Colantoni, A., Monarca, D. (2018). A combination of agricultural and energy purposes: Evaluation of a prototype of photovoltaic greenhouse tunnel. *Renewable and Sustainable Energy Reviews*, 82, 1178–1186.

Masini, E., Tomao, A., Barbati, A., Corona, P., Serra, P., Salvati, L. (2019).

Urban growth, land-use efficiency and local socioeconomic context: A comparative analysis of 417 metropolitan regions in Europe. *Environmental Management*, 633, 322–337.

Moretti, V., Salvati, L., Cecchini, M., Zambon, I. (2019). A long-term analysis of demographic processes, socioeconomic 'modernization' and forest expansion in a European country. *Sustainability* (Switzerland), 112, 388.

Moulaert, F., Rodríguez, A., Swyngedouw, E. (2003). *The Globalized City: Economic Restructuring and Social Polarization in European Cities*. Oxford: Oxford University Press.

Munafò, M., Norero, C., Sabbi, A., Salvati, L. (2010). Urban soil consumption in the growing city: A survey in Rome. *Scottish Geographical Journal*, 126(3), 153–161.

Munafò, M., Salvati, L., Zitti, M. (2013). Estimating soil sealing rate at national level—Italy as a case study. *Ecological Indicators*, 26, 137–140.

Muñoz, F. (2003). Lock living: Urban sprawl in Mediterranean cities. *Cities*, 20, 381–385.

Newman, P., Thornley, A. (2011). *Planning World Cities: Globalization and Urban Politics*. Macmillan International Higher Education.

Orenstein, D. E., Frenkel, A., Jahshan, F. (2014). Methodology matters: Measuring urban spatial development using alternative methods. *Environment and Planning B: Planning and Design*, 41, 3–23.

Pacione, M. (2003). Urban environmental quality and human wellbeing—a social geographical perspective. *Landscape and Urban Planning*, 65, 19–30.

Paul, V., Tonts, M. (2005). Containing urban sprawl: Trends in land use and spatial planning in the metropolitan region of Barcelona. *Journal of Environmental Planning and Management*, 48(1), 7–35.

Phelps, N. A., Parsons, N., Ballas, D., Dowling, A. (2006). *Post-suburban Europe: Planning and Politics at the Margins of Europe's Capital Cities*. Basingstoke: Palgrave MacMillan.

Pinson, D., Thomann, S. (2001). *La Maison et ses territories. In: De la villa a la ville diffuse*. Paris: L'Harmattan, Paris.

Portnov, B. A., Safriel, U. N. (2004). Combating desertification in the Negev: Dryland agriculture vs. dryland urbanization. *Journal of Arid Environments*, 56, 659–680.

Prokopová, M., Cudlín, O., Vèeláková, R., Lengyel, S., Salvati, L., Cudlín, P. (2018). Latent drivers of landscape transformation in eastern Europe: Past, present and future. *Sustainability* (Switzerland), 108, 2918.

Rekacewicz, P., Ahlenius, H. (2006). Coastal population and altered land cover in coastal zones (100 km of coastline). *Global Environment Outlook*, 4 (GEO-4).

Richardson, H. W., Chang-Hee C. B. (2004). *Urban Sprawl in Western Europe and the United States*. Aldershot: Ashgate.

Rontos, K., Grigoriadis, E., Sateriano, A., Syrmali, M., Vavouras, I., Salvati, L. (2016). Lost in protest, found in segregation: Divided cities in the light of the 2015 "Οχι" referendum in Greece. *City, Culture and Society*, 7(3), 139–148.

Rose, D., Harrison, E. (Eds.). (2014). *Social Class in Europe: An Introduction to the European Socio-economic Classification* (Vol. 10). Routledge, London.

Rosti, L., Chelli, F. (2012). Higher education in non-standard wage contracts. *Education and Training*, 54(2-3), 142-151.

Rubinfeld, D.L. (1987). The economics of the local public sector. In *Handbook of Public Economics*, 2, 571–645..

Sallustio, L., Pettenella, D., Merlini, P., Romano, R., Salvati, L., Marchetti, M., Corona, P. (2018). Assessing the economic marginality of agricultural lands in Italy to support land use planning. *Land Use Policy*, 76(2), 526–534.

Salvati, L. (2014). Towards a polycentric region? The socio-economic trajectory of Rome, an 'eternally Mediterranean' city. *Tijdschrift voor economische en sociale geografie*, 105(3), 268–284.

Salvati, L. (2019). Farmers and the city: Urban sprawl, socio-demographic polarization and land fragmentation in a mediterranean region, 1961–2009. *City, Culture and Society*, 18, 1002845.

Salvati, L., Colantoni, A. (2015). Land use dynamics and soil quality in agro-forest systems: A country-scale assessment in Italy. *Journal of Environmental Planning and Management*, 58(1), 175–188.

Salvati, L., Gargiulo Morelli, V. (2014). Unveiling urban sprawl in the Mediterranean region: Towards a latent urban transformation? *International Journal of Urban and Regional Research*, 38(6), 1935–1953.

Salvati, L., Guandalini, A., Carlucci, M., Chelli, F.M. (2017). An empirical assessment of human development through remote sensing: Evidences from Italy. *Ecological Indicators*, 78, 167–172.

Salvati, L., De Rosa, S. (2014). Hidden Polycentrism' or 'subtle dispersion'? Urban growth and long-term sub-center dynamics in three Mediterranean cities. *Land Use Policy,* 39, 233–243.

Salvati, L., Ferrara, A., Tombolini, I., Gemmiti, R., Colantoni, A., Perini, L. (2015). Desperately seeking sustainability: Urban shrinkage, land consumption and regional planning in a Mediterranean metropolitan area. *Sustainability*, 7(9), 11980–11997.

Salvati, L., Gargiulo Morelli, V., Rontos, K., Sabbi, A. (2013b). Latent exurban development: City expansion along the rural-to-urban gradient in growing and declining regions of southern Europe. *Urban Geography*, 34(3), 376–394.

Salvati, L., Sabbi, A. (2011). Exploring long-term land cover changes in an urban region of southern Europe. *International Journal of Sustainable Development and World Ecology*, 18(4), 273–282.

Salvati, L., Tombolini, I., Ippolito, A., Carlucci, M. (2018). Land quality and the city: Monitoring urban growth and land take in 76 southern European metropolitan areas. *Environment and Planning B: Urban Analytics and City Science*, 454(3), 691–712.

Salvati, L., Zambon, I., Chelli, F. M., Serra, P. (2018). Do spatial patterns of urbanization and land consumption reflect different socioeconomic contexts in Europe? *Science of The Total Environment*, 625, 722–730. Salvati, L., Zambon, I., Pignatti, G., Colantoni, A., Cividino, S., Perini, L., Pontuale, G., Cecchini, M. (2019). A time-series analysis of climate variability in urban and agricultural sites (Rome, Italy). *Agriculture (Switzerland)*, 95, 103.

Salvati, L., Zitti, M., Sateriano, A. (2013a). Changes in the city vertical profile as an indicator of sprawl: Evidence from a Mediterranean region. *Habitat International*, 38, 119–125.

Schneider, A., Woodcock, C. E. (2008). Compact, dispersed, fragmented, extensive? A comparison of urban growth in twenty-five global cities using remotely sensed data, pattern metrics and census information. *Urban Studies*, 45(3), 659–692.

Scott, A. J. (2001). Globalization and the rise of city-regions. *European Planning Studies*, 9(7), 813–826.

Seronde-Babonaux, A. (1983). *Roma: dalla città alla metropolis*. Bologna: Editori Riuniti.

Serra P., Saurí D., Salvati L. (2018). Peri-urban agriculture in Barcelona: Outlining landscape dynamics vis-à-vis socio-environmental functions. *Landscape Research*, 435(5), 613–631.

Tello, E., Ostos, J. R. (2012). Water consumption in Barcelona and its regional environmental imprint: A long-term history (1717–2008). *Regional Environmental Change*, 12(2), 347–361.

Terzi F., Bolen F. (2009). Urban sprawl measurement of Istanbul. *European Planning Studies*, 17(10), 1559–1570.

Tewdwr-Jones, M., McNeill, D. (2000). The politics of city-region planning and governance: Reconciling the national, regional and urban in the competing voices of institutional restructuring. *European Urban and Regional Studies*, 7(2), 119–134.

Theobald, D. (2001). Land use dynamics beyond the American urban fringe. *Geographical Review*, 91(3), 544–564.

Torrens, P. M. (2008). A tool kit for measuring sprawl. *Applied Spatial Analysis and Policy*, 1, 5–36.

Tsai, Y. H. (2005). Quantifying urban form: Compactness versus 'sprawl'. *Urban Studies*, 42(1), 141–161.

Turok, I., Mykhnenko, V. (2007). The trajectories of European cities, 1960–2005. *Cities*, 24(3), 165–182.

Vidal M., Domènech, E., Saurí, D. (2011). Changing geographies of water-related consumption: Residential swimming pools in suburban Barcelona. *Area*, 43(1), 67–75.

Wang, S., Yang, Z., Liu, H. (2011). Impact of urban economic openness on real estate prices: Evidence from thirty-five cities in China. *China Economic Review*, 22(1), 42–54.

Zambon, I., Cerdà, A., Cividino, S., Salvati, L. (2019). The (evolving) vineyard's age structure in the valencian community, Spain: A new demographic approach for rural development and landscape analysis. *Agriculture* (Switzerland), 93, 59.

Zambon, I., Rontos, K., Serra, P., Colantoni, A., Salvati, L. (2018). Population dynamics in southern Europe: A local-scale analysis, 1961–2011. *Sustainability* (Switzerland), 111, 109.

Zambon, I., Serra, P., Sauri, D., Carlucci, M., Salvati, L. (2017). Beyond the 'Mediterranean city': Socioeconomic disparities and urban sprawl in three southern European cities. *Geografiska Annaler: Series B, Human Geography*, 99(3), 319–337.

4

Management and Governance Implications for the "Suburban" Cities

Sabato Vinci

The Mediterranean region has a long and complex history, peculiar and unique characteristics, a wealth of the most bizarre and original urban morphologies shaped by a complex interplay of socioeconomic forces (Salvati, 2018; Cecchini et al., 2019; Carlucci et al., 2019). For these reasons, much has been said and written about the Mediterranean cities. According to Matvejevic (1998), everything has already been said of the Mediterranean from various disciplinary perspectives. Whether or not Matvejevic's statement is correct, the objective of this chapter is to investigate and discuss the most recent urban transformations in Southern Europe. The reasons underlying such an interest in the Mediterranean cities over other (transforming) centers of the world have several justifications. To begin with, Mediterranean cities have historically been considered as the prototype of the compact city (Carlucci et al., 2018). Nevertheless, most of these cities are gradually abandoning their compact tradition and switching to sprawl. Consequently, the analysis of urban sprawl is interesting and challenging in the Mediterranean context.

Furthermore, the Mediterranean region is passing through a period of uncertain transition. Therefore, studying the "territorial symptoms" of this undefined evolution will reveal useful to policy-makers for the adoption of correct strategic measures in order to preserve natural and human environments and, therefore, for the strategic management of metropolitan spaces (Bryson et al., 1983; Kovach at al. 1990; Salvati et al., 2017). Located in between the "global" cities of Northern Europe and the developing countries of the world, Mediterranean cities are aspiring to a stable position amongst the rich part of Europe. Nevertheless, urban structures deriving from past development determine inconsistencies and limitations to this desire. This mixture of forces that come from the past and weaknesses that consolidate in the present and that are projected in the future, are part of the genetic heritage of the Mediterranean urbanities. We argue that territorial transformations of the Mediterranean cities seem to crusade against the typical characteristics of this region. This has undoubtedly helped to understand the dynamics and causes of sprawl in a region which appears only apparently disordered and uncertain to the eyes of their inhabitants.

4.1 Past Dynamics and Future Trends

Urban dispersion is advancing in many metropolitan areas of the world, and it is common also in European cities, posing a governance problem for the new metropolitan areas, which takes into account economic, industrial, legal, and geomorphological factors (Parks et al., 1989; Gills et al., 1992; Briffault, 1996; Kübler et al., 2004). In fact, several Mediterranean cities have undergone a rapid transition from the traditional "compact" model to various steps of a more "dispersed" form characterized by an impressive sprawl around the urban area. Urban dispersion advances rapidly in Mediterranean Europe where urbanization expands at much faster rates than population growth in the most recent "deconcentration" wave. Examples of this trend can be found in Porto (European Environment Agency, 2006), Lisbon (Barradas and Salgueiro, 2003), Barcelona (Catalàn et al., 2004), Marseilles and the nearby Rhone valley (Pinson and Thomann, 2001), Milan (Camagni et al., 2002), Bologna (Anderlini, 2003), Rome (Munafò et al., 2013), Venice and the Veneto region (Indovina, 1990), Athens, and many other cities. In the report "Urban Sprawl in Europe" (2006), the European Environment Agency has found that 6 of the 10 European cities where sprawl is growing fastest are in Southern Europe (Istanbul, Palermo, Porto, Iraklion, Lyon, and Milan). Spanish cities do not escape this general trend: Madrid (hosting almost 5 million people in its metropolitan area) is perhaps the best example.

Taken together, the Mediterranean urban areas are experiencing a change toward more dispersed and horizontal rather than vertical growth at the expense of farming and forested areas, semi-natural environments, and wetlands (Brunori et al., 2018; Sallustio et al., 2018; Salvati et al., 2019; Zambon et al., 2019a, 2019b; Carlucci et al., 2019; Moretti et al., 2019). This trend serves to attenuate the levels of over-densification of central districts, but it poses also a problem of environmental sustainability in a public governance perspective. Indeed, on the one hand, there is a need to consistently expand the tangible and intangible connection networks, thus bearing all the costs derived from primary urbanization works. On the other hand, it also causes more homogeneous urban environments and the intrinsic standardization of rural landscapes (Ciommi et al., 2019). Earlier studies have demonstrated a correlation between intensified urbanization rates and an increase of natural disasters and hazards for resident populations.

4.1.1 Globalizing Economy, Destructuration of Traditional Rural Economies, and Coastal Urban Growth in Mediterranean Countries

Urbanization is caused by push-and-pull factors, the structure of the economy and the stage of economic development (Ciommi et al., 2018). Population growth has been a major driver for the rapid expansion of megacities and for informal housing that is highly vulnerable to disasters (Ciommi et al., 2017). Both past and projected population growth are crucial for assessing future vulnerability to disasters. In the Mediterranean basin, demographic data indicate two patterns of growth. Between 1850 and 2000, population in the five Southern European countries doubled, while that of the 12 European dialogue partners (*plus* Libya) increased 9-fold. From 2000 to 2050, a declining population has been projected in the five Southern and Southeastern European countries (except Albania), with slight increases in Cyprus, and major increases in North Africa (+97 million) and in the Eastern Mediterranean (+84.3 million). In the 12 Middle East and North African countries, more people will be added until 2050 than those presently living in the five Southern European countries (177.3 million).

Urbanization trends in the region have differed significantly. While in Southern Europe urbanization rate has been projected to increase by 2030, urbanization rate in northern Africa has been projected to grow even more rapidly. According to the United Nations Urbanization Prospects, by 2030, projections estimate that urban population will be 71.6% in Greece, 76.1% in Italy, 81.6% in Portugal, 82.2% in France, and 84.5% in Spain. These projections clearly evidence the growing pressure for urbanization that Southern European and Northern African cities will experience in the near future. In Southern Europe, from 1950 to 2000, the Mediterranean coastal cities (Athens, Barcelona, Naples, and Marseille) have experienced a gradual increase in urbanization rates (1.1–1.8 fold). Urbanization rates in these cities are expected to stabilize by 2015. Nevertheless, urbanization in the Mediterranean coastal cities is contradicted in some cases. For example, the capital of Turkey was experiencing a rapid increase in urbanization rates (10-fold during 1950–2000) and future predictions do not expect that these rates will stabilize or decrease.

Whilst more than half the world's population now lives in towns, two out of three inhabitants in the countries bordering the Mediterranean Sea already live in urban areas. In the Mediterranean coastal region, population went from 285 million in 1970 to 427 million in 2000 and will probably reach 524 million by 2025 according to Blue Plan scenarios. Furthermore, the same plan has predicted that urbanization rates will increase throughout the entire region. Urban growth is becoming increasingly endogenous, fed by internal redistribution of population, interurban migration and rural exodus, which is either drying up (Egypt, Tunisia, and Libya) or holding up (Turkey, Syria, and Morocco). Over one third of this growth will take place in coastal regions, more specifically in coastal cities. "Coastalisation" (the concentration of population and economic activities in coastal spaces) is (and will represent) a considerable source of urbanization in the Mediterranean region.

Coastalization of the Mediterranean shore has been a general trend since the last two centuries. This is due to the vast areas of hills and mountains that characterize the inland areas of the Mediterranean region, displaying considerable handicaps for urbanization. Coastalization has intensified in the last years of the 20th century due to growing international tourism on the Mediterranean shores. With 150 million tourists visiting the coastal regions, the Mediterranean basin is the primary tourist destination in the world, and the influx could double between now and 2025. Furthermore, the globalizing economy and the consequent destructuration of traditional rural economies and societies of the inland areas have significantly contributed to coastal urban expansion in the Mediterranean countries (Ciommi et al., 2018, 2019; Salvati et al., 2018). Coastalization of populations and economic activities has also been reinforced by major public works that littoral plains have developed, such as irrigation and drainage systems, demosquitoising and large-scale transport infrastructures.

Urban sprawl is becoming endemic along the European coastal regions. The environmental impact of sprawl is evident in several ecologically sensitive areas located in coastal districts and mountain areas. The Plan Bleu and Center d'Activités Régionales (2001) assumed a predicted increase in population within a maximum figure of 35 million people. Development-related impacts on coastal ecosystems, and their habitats and services, have produced major changes in these zones. The Mediterranean seacoast, one of the world's 34 biodiversity hotspots, is particularly affected, and the increased demand for water for urban use competes with irrigation water for agricultural land.

This problem has been exacerbated by the increased development of golf courses in Spain, where the over-extraction of groundwater has led to saltwater intrusion into the groundwater. The consequences of urban sprawl on the environment, economy, and quality of cities include:

- ever-growing cost of urban infrastructures;
- degradation of urban lifestyles (especially related to the problem of urban mobility);
- loss of farming industry and natural land along the seacoasts;
- disappearance of wetlands and coastal erosion;
- destruction of highly valuable natural habitats (shallow water areas, posidonia beds, sand dunes, and turtle nesting sites);
- reduction of small scale fishing industry;
- global degradation of the Mediterranean landscapes and
- an increased risk of natural disasters in urban areas.

The list above refers specifically to the consequences of urban sprawl along the Mediterranean coasts. This list can be further expanded if the observed impacts of low-density urbanization in the rest of the region are considered, including:

- a constant increase in demand for travel, linked with decoupling of home from work, with an increase of public and private transportation costs;
- generalized congestion along the main transport roads and, consequently, a drop-in travel speed;
- recurrent shortcomings in public transport provision in terms of services' levels;
- a constant rise in greenhouse gas emissions linked with the transport sector, mainly road transport, which is heavily dependent on fossil energy;
- increased per-capita land consumption;
- higher dependency on fossil fuels, and
- loss of economic attractiveness characteristic of central city.

Faced with these trends, policies of integrated management of coastal areas, sustainable conservation of the coastline, wetlands and peri-urban farming land, should be strengthened with the final aim at promoting sustainable agriculture and rural development of inland areas, and integrating tourism and sustainable development (Fabbio et al., 2018; Marchi et al., 2018; Masini et al., 2019; Bertini et al., 2019). The Mediterranean

Commission on Sustainable Development endeavors to produce ideas and strategic proposals in this regard, and efforts are being made at regional levels in most of the Mediterranean countries, but empirical results are still limited.

4.2 Climate Change, Environmental and Public Health Emergencies

According to the Third Assessment Report (TAR) of the Intergovernmental Panel on Climate Change (IPCC, 2001), between 1990 and 2100, "the globally averaged surface temperature is projected to increase by 1.4?C to 5.8?C and the mean sea level will rise by 0.09 to 0.88 meters." The TAR argued that global climate change increased the probability of some extreme weather events during the 20th century and that, in the 21st century, "more intense precipitation events" and an "increase of the heat index" will become "very likely, over most areas." Due to regional climate differences, "expected climate change will give rise to different exposures to climate stimuli across regions". The IPCC concluded that less-developed regions are severely vulnerable and "in Europe, vulnerability is significantly greater in the South."

This has also been stressed in the IPCC assessments of the climate scenarios for Europe pertaining to changes in temperature and precipitation during summer for the 2020s, 2050s, and 2080s. Projected trends for Southern Europe are obvious: air temperature may increase, and precipitation may decline mostly in the Mediterranean countries. Climate change produces short-term and long-term socioeconomic impacts that can contribute to disasters that vary according to the specific vulnerability, that may be reduced by adaptation and mitigation measures (Salvati et al., 2009, 2011, 2013; Colantoni et al., 2015; Salvati et al., 2019; Proietti et al., 2019; Francaviglia et al., 2019). In response to human activities and the natural environment, Europe and the Mediterranean area are severely exposed to the risk to redefine their traditional industrial activities, and the public administrations themselves are faced with new problems of managing an orderly development of urban areas. From the economic point of view, industrial activities mainly suffer the sensitive impact of extreme seasons (exceptionally hot and dry summers, mild winters). Instead, from the governance point of view, short-duration hazards (windstorm, heavy rain, river valley flooding) and long-term changes (coastal squeeze, sea-level rise) claim for new public organizational tools, both for natural disasters management and redefinition of territorial structures. In fact, climate changes will impact human settlements, worsening the current trends in anthropogenic pressure. According to the IPCC (2001), "in such areas, squatter and other informal urban settlements with high population density, poor shelter, little or no access to resources such as safe water and public health services and low adaptive capacity are highly vulnerable". In fact, the expected climate change was associated with the increase in heat waves and the rise in floods: so, it "will increase the risk of drowning, diarrhoea and respiratory diseases, and in developing countries, hunger and malnutrition" (IPCC, 2001). Therefore, the local health authorities are called to contribute to the public response to such changes in terms of both preventive/public health and caring (Ebi et al., 2006; Wilson, 2006; Frumkin et al., 2008; Keim, 2008).

It can be argued that the impact of natural emergencies (therefore, also enviromental and health type emergencies) and the climate change impact on the economic and social life of local communities is determined by hazards whose intensity is influenced by social, economic, physical, and environmental vulnerabilities (Yole, 1990; Cutter, 2003; Dobó, 2006; Naudé, 2009; World Health Organization, 2013). In this sense, with particular reference to the relation between environmental emergencies and urban structure, it appears that poorly constructed un-planned settlements of Mediterranean cities are the most vulnerable. In particular, the overall vulnerability of the region is the highest in Europe, because spontaneous suburban neighborhoods are most common in these types of cities.

Geophysical and hydro-meteorological disasters share common features, but natural and human-induced environmental challenges impact the Mediterranean region differently. While climate change, desertification, and the hydrological cycle have contributed to environmental degradation during the 20th century (Biasi et al., 2019; Francaviglia et al., 2019; Proietti et al., 2019; Salvati et al., 2019), human-induced factors (population growth, urbanization, and agriculture/food demands) will increase pressure on the environment even more during the 21st century. These trends have impacted the vulnerability of urban centers to natural disasters in Europe, the Middle East, and North Africa.

Both international organizations and individual experts noted the increasing disaster potential of megacities. Mitchell (1999) noted that megacity hazards, such as floods, earthquakes, and windstorms, are the most common shocks, followed by other risks that trigger disasters. In this regard, Mitchell noted major changes in megacity hazards but also in interactivity, risks, change in exposure, vulnerability, and the efficacy of hazard management. Not only vulnerability level will rise, but also the impact of extreme weather events. A survey of natural disasters from 1975 to 2001 in the Mediterranean basin indicates that more than half of all the natural disasters were reported for the five South European countries. The survey contributed to demonstrate which urban centers show the highest vulnerability against natural disasters. Most natural disasters were reported in France (86), followed by Turkey (63), Italy (57), Spain (47), Greece (43), Algeria (36), and Morocco (23). Regarding fatalities, Turkey ranked first (27,375), followed by Italy (6,158) Algeria (4,124), Greece (1,573), and Egypt (1,386). But, for affected persons, Spain was in the lead (6,819,987), followed by France (3,890,759), Albania (3,259,759), Turkey (2,580,392), and Algeria (1,154,355).

Around the Mediterranean, most disaster fatalities were the result of earthquakes (Turkey, Italy, and Algeria), although drought and famine affected more people. Before the 1980s and the 1990s, disaster fatalities increased for Turkey, Egypt, Morocco, Italy, and France, while the affected persons increased most for Spain but also for France. In Northern Africa, the number of people affected by natural disasters increased from the 1980s to the 1990s for Egypt, Morocco, and Algeria, while the numbers declined for Tunisia: 96,000 inhabitants in 1970, 30,000 in 1982, and 2,500 in 1986. Between 1975 and 2001, geophysical disasters caused most fatalities (in Turkey, Italy, and Algeria), while the winter storms in France, and drought in Spain, Albania, Syria, and Morocco, caused an increasing number of disaster-affected people (Incerti et al., 2007; Salvati et al., 2009; Salvati and Colantoni, 2015; Zambon et al., 2017). The empirical results of this survey evidence how rapid urbanization has increased and will further increase the vulnerability to all types of disasters, especially for the poor living in informal housing and in flood-prone areas. This result is of particular concern as the probability and intensity of hydrometeorological disasters have been projected to increase because of climate change (Biasi et al., 2019). Consequently, rapid urbanization may further increase vulnerability to disasters and fatalities in the years to come.

An additional phenomenon concerns the public health emergencies (mainly epidemics), which have the ability to spread in densely populated territories and urban areas. Italy is taken as a representative case for this topic. Italy is not only a highly urbanized and densely populated country, but it was also the first Country massively affected in Europe by the Covid-19 epidemic (Spina, 2020), that caused a lot of social tensions and a real humanitarian and economic nightmare (*sensu* McKibbin, 2006; Verikios,

2011; Peckham, 2013). In this regard, the recent Italian health experience constitutes a precious opportunity to start a brief reflection about the critical points of the Italian emergency management system, taken as a paradigmatic example of preparedness (or non-preparedness) to risks and disasters typical of Mediterranean countries.

The national reforms of the 1990s (Legislative Decree no. 502 dated 30 December 1992; Legislative Decree no. 517 dated 7 December 1993; Legislative Decree no. 229 dated 19 June 1999) encouraged the opening up of the National Health System to the free market, based on the assumption to separate public companies that purchase health services for citizens (Local Health Authorities) and those producing such services, including Hospital Authorities and private individuals "accredited" with the NHS (Caruso, 2009). In other words, the ambition of the system was to proceed with a partial "disintegration" of the three pillars of public health (preventive medicine, hospital care, emergency medicine) that were integrated into the pre-1992 health model. In particular, the ambition was to keep the first pillar (preventive medicine) and the third pillar (emergency medicine) in the hands of the local health authorities and to open the second pillar (hospital activity) to a form of competition "administered" by public bodies (Compagnoni, 2006).

The basic theory was grounded on the "Quasi-Market" assumption (Jones and Cullis, 1996; Kitchener, 1998; Nicita, 2004; Hills & Le Grand, 2007; Petretto, 2010; Carabelli & Facchini, 2011), i.e. a model in which basically the Local Health Authorities act as "third payer" (instead of the citizen) and the citizen can choose indifferently whether to be treated in public or private "accredited" hospitals. In both cases, the medical-healthcare service is reimbursed to the provider, on the basis of a standard cost system, namely the Diagnosis-Related Group, DRG (Nonis and Rossi, 2009), provided that it is consistent with the public programming established on the basis of the volume of financial resources allocated by the respective administrative region - which in turn receive them from the State on the basis of a national breakdown.

The "Quasi-Market" system might seem efficient, as it should bring public and private hospitals to compete with each other to win the trust of the patient who chooses whom to turn to. In reality, it was the Italian region that applied this theory in the most rigorous way (Lombardy, Northern Italy) that proved to be the least equipped to handle public health emergencies such as Covid-19. In fact, almost all Italian regions have suffered little from attempts to open up to competition, choosing to maintain an health system essentially public and integrated between the three pillars, only corrected by the system of the "closed list" of private operators accredited to the NHS (the weight of private health care in Italy is, on average, under 6% of total expenditure). Conversely, Lombardy has chosen a very different path compared with the other Italian regions, creating a highly competitive system between public and private hospital structures (similar to the health care systems of other countries such as the Netherlands, the United Kingdom, Germany, Sweden) and thus implementing, almost faithful to the basic theory, the Quasi-Market project. However, this model, which opened to competition between public and private structures the most remunerative pillar of the entire health care (the hospital care), presented the "side effect" of encouraging the concentration of resources and political-administrative interests (removing them, above all, from the activity of prevention and territorial medicine), thus orienting the system to the acquisition of patients to be treated. In fact, the payment of the services provided by both public and private hospital facilities by the acquiring public health authority is based on DRGs and therefore on the presence of patients to be treated. In the specific case of Lombardy, there have been, basically, two phenomena especially in the last decade:

a. most of the regional health expenditure has been allocated to the pillar of diagnosis and treatment, taking away resources and attention especially to prevention and therefore to territorial medicine;
b. Lombardy healthcare system, focusing above all on excellent hospitalization, invested in attracting patients from other Italian regions (especially from Southern Italy) whose healthcare services, through intra-regional reimbursement, are paid to Lombardy by the regions of origin.

The overall "hospital-focused" vocation of Lombardy seems to have was demonstrated to be the weak point in the fight against Covid-19, whose efficient management would have required a stronger public health structure and a network of general practitioners put in a position to effectively and autonomously monitor and manage patients, instead of a thorough policy of hospitalization. In fact, the two regions that had started with a similar intensity of contagion - Lombardy and Veneto (which are very similar, both culturally and politically) - responded in a different way to the pandemic, having opposite results in the medium-term. Lombardy, leveraging on its "hospital-focused" cultural model, favoured extensive hospitalization, with the side-effect of spreading the contagion even among the health workers themselves and within the health structures, and this has produced the further spread of the contagion. Veneto, whose health system has remained more faithful to the original model (integration between the prevention, hospital care, and emergency pillars), provided tools to counter the pandemic: preventive medicine actions, territorial medicine network and home care. In this way, Veneto region has had much better results than the neighbouring Lombardy region. However, what is most important to understand the differences between healthcare in Lombardy and the rest of Italy, is the cultural model of reference, which has concrete consequences also on the non-economic components on which a healthcare model is based. For example, despite the described "Quasi-Market" system and the consequent incentive to concentrate investments mainly on the treatment segment (sacrificing above all prevention and territorial medicine), general practitioners are also present in Lombardy by national law, exactly according to the same rules and the same physician-inhabitant relationship provided for the rest of Italy. The fact that they have not managed to effectively monitor the territory, probably depends on the assumption that most regional policy-makers and healthcare administrators on the hospital segment have been accompanied by the overshadowing of activities such as coordination and controls.

However, in general (with the exception of Lombardy), it can be said that the Italian healthcare system has responded relatively well to the pandemic, without presenting any particular shortcoming from the point of view of emergency health management capacity. On the contrary, the Italian healthcare model shown great flexibility, for example, expressed in its ability to significantly and rapidly increase the number of beds in intensive care and to organize structures specialized in the treatment of Covid-19. With particular reference to the intensive care units, it should be pointed out that the country's ordinary budget is set against "ordinary" situations, and compared to these, it is quite adequate. In fact, Italy ordinarily has 5,300 intensive and sub-intensive care places, divided between public (70%) and private (30%) hospitals: this corresponds to 13.5 beds per 100 thousand inhabitants, equal to about 3.3% of the total number of beds used for acute patients. There are also European countries with a greater endowment: for example, Germany (29.2 beds for every 100 thousand inhabitants) and Austria (21.8 beds for every 100 thousand inhabitants). However, according to data published in 2012 in one of the most authoritative scientific journals in the sector (Intensive Care Medicine), there are also countries with a lower budget than Italy: the Scandinavian countries, the Netherlands, the United Kingdom, as well as Spain. Among other things, as argued in part of the both medical and health management scientific debate, the number of places in intensive care is not an independent variable in health care. In fact, the number of intensive care posts should be related to the overall level of the system, as, for example, fewer posts may ordinarily be needed in a more advanced system than those considered adequate in a less advanced system.

The problems that have emerged in some areas of emergency management thus seem to be essentially "context-specific". On the one hand, it was noted the unfamiliarity of some decision-makers with decision-making instruments formally provided for (in the legislation) but scarcely used (local, regional, national restrictive ordinances, doubts about spheres of competence). On the other hand, the lack of familiarity cannot be too surprising, if we consider the Covid-19 emergency was the first real national health emergency in Italy since the end of the Second World War. The intrinsic difficulty of the Italian industry to satisfy autonomously the extraordinary demand for health devices added in a global pandemic scenario, in which many countries have requisitioned such devices on their national territory, blocking or hindering exports. This demonstrates the importance of having a national industry in strategic sectors broader than those considered in the 1990s, in order to ensure the self-sufficiency of national political-administrative systems (which continue to be the main health reference) in the face of disasters and systemic emergencies.

For the rest, the Italian health sector does not seem to have particular structural problems to face. So, it can be argued overall that Italy has an excellent universalistic National Health Service, strongly rooted in the territory, which allowed management of prevention actions and treatment of infected people, albeit with the specificity of the Lombardy health system. In addition, the Italian government has been able to create public order measures (until the temporary prohibition for citizens to leave their home except for urgent work, health or other reasons) in order to contain the virus diffusion, mainly in urban areas. With reference to these elements, Italy has demonstrated an overall adequate system: for instance, it is possible to imagine that non-universalistic health systems would have greater difficulties in handling health emergencies, as a serious pandemic (Ugolini, 2004). However, such emergencies have also shown that there are still some margins of improvement in the governance of Italian health and emergency system, with particular regard to the organizational and administrative point of view. Therefore, a brief general reflection may be useful in this regard.

First, it emerges the opportunity of substantial strengthening of the chain of command, working on the coordination of a very fragmented decision-making process between a multiplicity of central and local bodies and governments. In fact, a too fragmented chain of command can be an obstacle to the speed of execution, reaction and effectiveness of democratic and administrative structures with respect to the speed and risks of the contemporary world. Indeed, the management of a natural emergency requires the ability of the public administration to implement rapid and incisive public governance responses, by coordinated actions of all the levers of the social organization: national government, local governments, law enforcement, transport system of men and means, healthcare, logistics and the supply system.

Second, the Italian governance system would be much more fluid, operating one radical procedural simplification in the processes of purchase of Public Administration. The Code of Public Contracts in Italy was reformed with a Legislative Decree dated 18 April 2016, no. 50, and to this legislation was added a "corrective decree" by Legislative Decree dated 19 April 2017, no. 56, plus further amendments provided by a national Law dated 14 June 2019, no. 55). Numerous side rules were added: 62 implementing acts, including decrees and guidelines of the Italian National Anti-Corruption Authority (ANAC), most of which have not yet been issued. Therefore, the Italian public procurement system constitutes an extremely complex world for economic operators and public decision makers alike. Moreover, the Italian public procurement legislation appears to be dominated by the anti-corruption logic (it emerges from the main role assumed by ANAC) than by the logic of administrative efficiency: so, it discourages public administrators from taking decision-making responsibilities. The introduction of this public authority was the most important change in recent years, which represents the opposite of the development of Italian history (Cassese, 2017, p. 64).

In fact, not only with reference to the management of public contracts related to the emergency from Covid-19, but more generally with reference to the problem of relaunching the infrastructural sector in Italy, the 2020 political debate seemed oriented towards the preparation of public procurement management procedures by way of derogation from ordinary legislation. The Italian President of the Council of Ministers in 2020, even spoke of a "Morandi bridge model", with reference to the well-known event of urgent reconstruction of the Italian bridge collapsed in Genoa on August 14, 2018. This reconstruction took place, precisely, in derogation from the public procurement ordinary legislation. So, the Italian policymakers consider that ordinary legislation is leading to excessive slowdowns on a procedural and managerial level. From this point of view, Decree Law no. 76 dated 16 July 2020 (the so-called "simplifications" Decree) has put in place a series of legislative provisions, in derogation from the Italian Public Contracts Code, aimed at encouraging public investments in the infrastructure and public services sector, dealing with the negative economic repercussions following the containment measures and the global health emergency of the COVID-19. These innovations, although having a temporary value (currently until July 31, 2021), are certainly to be welcomed, even if the decision to make the awarding of public contracts "below the threshold" (i.e. $< 150{,}000$ euros) practically discretionary for public administrators should be well considered. In particular, what seems essential is to achieve a virtuous balance between the opportunity to simplify procedures and the need not to weaken the fundamental rules upholding a good administration.

More generally, political decision-makers should consider as their primary objective a structural reform of public administration, starting from the idea that the public sector can "create value" as well as and even more than private agents (Chelli et al., 2016). The key point of the public administration reform in Italy was to find a balance between legal procedures and a greater focus on managerial issues, understood as the ability to "do things in the most effective ways and with the least use of resources allowed by the application of knowledge about the governance of complex systems" (Lega, 2016). An integral part of this renewal process should also be an increased focus on specific training for a new public management (Gigliarano and Chelli, 2016). In particular, considerable institutional incentives should be put in place to motivate those who are already in public administrative careers to acquire managerial skills, but above all it implies the ability to design specific higher education paths for those who will access them in the coming years (Chelli and Rosti, 2002; Rosti and Chelli, 2009; Chelli et al., 2009). These are indispensable things for modern public governance, capable not only of ordinarily administering public interests in an efficient way, but also of managing emergencies that have an impact on the territory and local communities, governing the most appropriate public response (Rosti and Chelli, 2012).

A further strategic element of the emergency management system (which has been evident in the Italian experience of fighting the Covid-19 virus) concerns the industrial problem. Therefore, it seems appropriate to make a brief reference to it. In fact, the Covid-19 virus has created a real global health emergency: this is what experts call "pandemic" (Gostin, 2004; Hayden, 2006; Bell et al., 2009; French et al., 2009; Vaughan et al., 2009; Kelly, 2011). On Jan 30, 2020, WHO declared the COVID-19 outbreak a public health emergency of international concern. In this context, the governments of many countries, where companies supplied Italy with medical materials (e.g. personal protective equipments), requisitioned these devices on their national territory and/or blocked these exports. This has shown the importance of having a national industry in strategic sectors larger than those considered during the 1990s: in fact, it is important to guarantee the self-sufficiency of national political-administrative systems - which continue to be the main reference of the populations in order emergency response - with regards at least the basic tools to deal with catastrophes and systemic emergencies (Lamonica and Chelli, 2018).

Therefore, it emerges how natural disasters and emergencies are not exclusively "technical" problems (environmental, engineering or health type). In fact, the emergency management, in the interest of the affected communities, requires a public administration system capable of governing the many variables that make up the problems – social, technical, legal, administrative, economic, logistic – by setting rapid and efficient responses.

4.3 Conclusions

The growing pressure on the environment due to human-induced factors (urbanization and natural resources consumption) is increasing the potential of disasters on megacities, creating new social problems. Cities are expanding into wider regions, thus increasing their exposure to natural changes (and natural disasters) and having a profound impact on both the city economy and lifestyles. In particular, illegal settlements contribute to the high vulnerability that the Mediterranean has been awarded with (Salvati et al., 2009, 2011, 2013; Colantoni et al., 2015). Furthermore, this situation is raising major concerns, as extreme weather events are likely to increase in frequency and intensity due to climate changes (Biasi et al., 2019). The growing vulnerability of urban centers along the Mediterranean coasts is testified by the increasing damages related to natural hazards. A survey on natural disasters in the Mediterranean basin shows that the region is more vulnerable to extreme seasons, short-duration hazards, such as floods and earthquakes, and slower long-term changes, such as sea-level rise and coastal squeeze. More in general, the new emergencies seem to have originated from climate changes and environmental or health phenomena (i.e. new pandemic viruses). These would be associated with events as the increase of the heat waves, the rise of floods and earthquakes (Kapucu; 2006; Bullock et al., 2017; Sylves, 2019), as well as old and new problems of public health, including diarrhoea and respiratory diseases and, in developing countries, hunger and malnutrition (Waring et al., 2005; Frumkin, 2008).

Thus, the challenge of climate change and environmental problems, in the peculiar urban context of Europe and the Mediterranean area, implies careful assessment by public authorities about their economic and social effects and about the most appropriate strategies of public governance. In fact, the emergency management in the interest of the affected communities, requires a public administration system that is capable of governing many variables (social, technical, legal, administrative, economic, logistic) together, combining different public powers (central government, local governments, law enforcement, health authorities) in a single strategy, so setting rapid and efficient public response.

References

Allen, P. (1995). Contracts in the National Health Service internal market. *The Modern Law Review*, 58(3), 321-342.

Allen, P. (2013). An economic analysis of the limits of market based reforms in the English NHS. *BMC Health Services Research*, 13(1), 1-10.

Alphan, H. (2003). Land use change and urbanization of Adana, Turkey. *Land Degradation and Development*, 14(6), 575–586.

Antrop, M. (2004) Landscape change and the urbanization process in Europe. *Landscape and Urban Planning*, 67(1-4), 9–26.

Bartlett, W. (1991). Quasi-markets and contracts: A markets and hierarchies perspective on NHS reform. *Public Money & Management*, 11(3), 53-61.

Bell, D.M., Weisfuse, I.B., Hernandez-Avila, M., Del Rio, C., Bustamante, X., Rodier, G. (2009). Pandemic influenza as 21st century urban public health crisis. *Emerging Infectious Diseases*, 15(12), 1963.

Beriatos, E., Gospodini, A. (2004). "Glocalising" urban landscapes: Athens and the 2004 olympicsOlympics. *Cities*, 21, 187–202.

Bertini, G., Becagli, C., Chiavetta, U., Ferretti, F., Fabbio, G., Salvati, L. (2019). Exploratory analysis of structural diversity indicators at stand level in three Italian beech sites and implications for sustainable forest management. 2019, *Journal of Forestry Research*, 301, 121–127,

Biasi, R., Brunori, E., Ferrara, C., Salvati, L. (2019). Assessing impacts of climate change on phenology and quality traits of *Vitis vinifera* L.: The contribution of local knowledge. 2019, *Plants*, 85, 121.

Briassoulis, H. (2004). The institutional complexity of environmental policy and planning problems: The example of Mediterranean desertification. *Journal of Environmental Planning and Management*, 47(1), 115–135.

Briffault, R. (1996). The local government boundary problem in metropolitan areas. *Stanford Law Review*, 1115–1171.

Bruegmann R. (2005). *Sprawl: A compact Compact historyHistory*. Chicago: University of Chicago Press, Chicago.

Brunori, E., Salvati, L., Antogiovanni, A., Biasi, R. (2018). Worrying about ''vertical landscapes": Terraced olive groves and ecosystem services in marginal land in central Italy. *Sustainability* (Switzerland), 104(2), 1164, 2.

Bryson, J.M., Boal, K.B. (1983). *Strategic Management in a Metropolitan Area. In Academy of Management Proceedings*, Vol. 1983, No. 1, 332–336. Briarcliff Manor, NY 10510: Academy of Management.

Bullock, J.A., Haddow, G.D., Coppola, D.P. (2017). *Introduction to Emergency Management*. Butterworth–Heinemann.

Carabelli, G., Facchini, C. (2010). *Il modello lombardo di welfare*. Milano: Franco Angeli.

Carlucci, M., Chelli, F.M., Salvati, L. (2018). Toward a new cycle: Short-term population dynamics, gentrification, and re-urbanization of Milan (Italy). *Sustainability (Switzerland)*, 10(9), 3014.

Carlucci, M., Zambon, I., Colantoni, A., Salvati, L. (2019). Socioeconomic development, demographic dynamics and forest fires in Italy, 1961–2017: A time-series analysis. *Sustainability* (Switzerland), 115, 1305.

Cassese, S. (2017). Se la politica soffoca le politiche. *Mondoperaio*, 5, 61–64.

Catalàn, B., Sauri, D., Serra, P. (2008). Urban sprawl in the Mediterranean? Patterns of growth and change in the Barcelona Metropolitan metropolitan Region region 1993–2000. *Landscape and Urban Planning*, 85(3–4), 174–184.

Cecchini, M., Cividino, S., Turco, R., Salvati, L. (2019). Population age structure, complex socio-demographic systems and resilience potential: A spatio-temporal, evenness-based approach. *Sustainability* (Switzerland), 117, 2050.

Chelli, F., Rosti, L. (2002). Age and gender differences in Italian workers' mobility. *International Journal of Manpower*, 23(4), 313-325.

Chelli, F., Gigliarano, C., Mattioli, E. (2009). The impact of inflation on heterogeneous groups of households: An application to Italy. *Economics Bulletin*, 29(2), 1276-1295.

Chelli, F.M., Ciommi, M., Emili, A., Gigliarano, C., Taralli, S. (2016). Assessing the Equitable and Sustainable Well-Being of the Italian Provinces. *International Journal of Uncertainty, Fuzziness and Knowlege-Based Systems*, 24, 39-62.

Chorianopoulos, I., Pagonis, T., Koukoulas, S., Drymoniti, S. (2010). Planning, competitiveness and sprawl in the Mediterranean city: The case of Athens. *Cities*, 27(4), 249–259.

Ciommi, M., Gigliarano, C., Emili, A., Taralli, S., Chelli, F.M. (2017). A new class of composite indicators for measuring well-being at the local level: An application to the Equitable and Sustainable Well-being (BES) of the Italian Provinces. *Ecological Indicators*, 76, 281-296.

Ciommi, M., Chelli, F.M., Carlucci, M., Salvati, L. (2018). Urban growth and demographic dynamics in southern Europe: Toward a new statistical approach to regional science. *Sustainability (Switzerland)* 10(8), 2765.

Ciommi, M., Chelli, F.M., Salvati, L. (2019). Integrating parametric and non-parametric multivariate analysis of urban growth and commuting patterns in a European metropolitan area. *Quality and Quantity*, 532, 957–979.

Colantoni, A., Ferrara, C., Perini, L., Salvati, L. (2015). Assessing trends in climate aridity and vulnerability to soil degradation in Italy. *Ecological Indicators*, 48, 599–604.

Compagnoni, V. (2006). Modelli di concorrenza e riforme sanitarie. *Economia Pubblica*, 3, 105-137.

Costa, F., Noble, A. G., Pendleton, G. (1991). Evolving planning systems in Madrid, Rome, and Athens. *Geojournal*, 24(3), 293–303.

Couch, C., Petschel-held, G., Leontidou, L. (2007). *Urban Sprawl in Europe: Landscapes, Land-use Change and Policy*. London: Blackwell.

Cutter, S.L., Boruff, B.J., Shirley, W.L. (2003). Social vulnerability to environmental hazards. *Social Science Quarterly*, 84(2), 242–261.

Davoudi, S. (2003). European briefing: polycentricity Polycentricity in European spatial planning: from an analytical tool to a normative agenda. *European Planning Studies,* 11(8), 979–999.

Di Novi, C., Piacenza, M., Robone, S., & Turati, G. (2019). Does fiscal decentralization affect regional disparities in health? Quasi-experimental evidence from Italy. *Regional Science and Urban Economics*, 78, 103465.

Dobó, E., Fekete-Farkas, M., Kumar Singh, M., Szûcs, I. (2006).

Ecological-economic analysis of climate change on food system and agricultural vulnerability: a A brief overview. *Cereal Research Communications*, 34(1), 777–780.

Dura-Guimera, A. (2003). Population deconcentration and social restructuring in Barcelona, a European Mediterranean city. *Cities*, 20(6), 387–394.

Ebi, K.L., Kovats, R.S., Menne, B. (2006). An approach for assessing human health vulnerability and public health interventions to adapt to climate change. *Environmental Health Perspectives*, 114(12), 1930–1934.

European Environment Agency (2006). *Urban sprawl in Europe – The ignored challenge*. Copenhagen: EEA Report no No. 10.

European Environment Agency (2010). *Mapping guide for a European urban atlas*. Copenhagen: Version 1.1. EEA.

Exworthy, M., Powell, M., Mohan, J. (1999). Markets, bureaucracy and public management: the NHS: quasi-market, quasi-hierarchy and quasi-network? *Public Money & Management*, 19(4), 15-22.

Fabbio, G., Cantiani, P., Ferretti, F., Di Salvatore, U., Bertini, G., Becagli, C., Chiavetta, U., Marchi, M., Salvati, L. (2018). Sustainable land management, adaptive silviculture, and new forest challenges: Evidence from a latitudinal gradient in Italy. *Sustainability* (Switzerland), 10(7), 2520.

Faludi, A.K.F. (2006) From European spatial development to territorial cohesion policy. *Regional Studies*, 40(6), 667–678.

Francaviglia, R., Di Bene, C., Farina, R., Salvati, L., Vicente-Vicente, J.L. (2019). Assessing "4 per 1000" soil organic carbon storage rates under Mediterranean climate: A comprehensive data analysis. *Mitigation and Adaptation Strategies for Global Change*.

French, P.E., Raymond, E.S. (2009). Pandemic influenza planning: An extraordinary ethical dilemma for local government officials. *Public Administration Review*, 69(5), 823–830.

Frumkin, H., Hess, J., Luber, G., Malilay, J., and McGeehin, M. (2008).

Climate change: The public health response. *American journal Journal of Public Health*, 98(3), 435–445.

Genske, D. D. (2003). *Urban Land – Degradation, Investigation, Remediation*. Berlin: Springer.

Giannakourou, G. (2005). Transforming spatial planning policy in Mediterranean countries: Europeanization and domestic change. *European Planning Studies*, 13(2), 319–331.

Gigliarano, C., Chelli, F.M. (2016). Measuring inter-temporal intragenerational mobility: an application to the Italian labour market. *Quality and Quantity*, 50(1), 89-102.

Gill, S. E., Handley, J. F., Roland Ennos, A., Pauleit, S., Theuray, N., Lindley, S. J. (2008). Characterising the urban environment of UK cities and towns: A template for landscape planning. *Landscape and Urban Planning*, 87(3), 210–222.

Gillis, M., Perkins, D. H., Roemer, M., Snodgrass, D. R. (1992). *Economics of Development*. WW Norton & Company, Inc.

Gospodini, A. (2009). Post-industrial trajectories of Mediterranean European cities: the case of post-Olympics Athens. *Urban Studies*, 46(5–6), 1157–1186.

Gostin, L. O. (2004). Pandemic influenza: public health preparedness for the next global health emergency. *The Journal of Law, Medicine & Ethics*, 32(4), 565–573.

Hamilton, D. K. (2000). Organizing government structure and governance functions in metropolitan areas in response to growth and change: A critical overview. *Journal of Urban Affairs*, 22(1), 65–84.

Hasse, J. E., Lathrop, R. G. (2003). Land resource impact indicators of urban sprawl. *Applied Geography*, 23(2–3), 159–175.

Hayden, F. G. (2006). Antiviral resistance in influenza viruses—implications for management and pandemic response. *New England Journal of Medicine*, 354(8), 785–788.

Herrschel, T. (2009). City regions, polycentricity and the construction of peripheralities through governance. *Urban Research and Practice*, 2(3), 240–250.

Hills, J., Le Grand, J. (2007). *Making social policy work*. Policy Press.

Incerti, G., Feoli, E., Salvati, L., Brunetti, A., and Giovacchini, A. (2007).

Analysis of bioclimatic time series and their neural network-based classification to characterise drought risk patterns in South Italy. *International Journal of Biometeorology*, 51(4), 253–263.

Ioannidis, C., Psaltis, C., Potsiou, C. (2009). Towards a strategy for control of suburban informal buildings through automatic change detection. *Computers, Environment and Urban Systems*, 33(1), 64–74.

Johnson, D. L., Lewis, L. A. (2007). *Land Degradation – Creation and Destruction*. Lahnam: Rowman and Littlefield.

Jomaa, I., Auda, Y., Abi Saleh, B., Hamzé, M., Safi, S. (2008). Landscape spatial dynamics over 38 years under natural and anthropogenic pressures in Mount Lebanon. *Landscape and Urban Planning*, 87(1), 67–75.

Kahn, M. E. (2000). The environmental impact of suburbanization, *Journal of Policy Analysis and Management*, 19(4), 569–586.

Kapucu, N. (2006). Interagency communication networks during emergencies: Boundary spanners in multiagency coordination. *The American Review of Public Administration*, 36(2), 207–225.

Kasanko, M., Barredo, J. I., Lavalle, C., McCormick, N., Demicheli, L., Sagris, V., Brezger, A. (2006). Are European Cities Becoming Dispersed? A Comparative Analysis of Fifteen European Urban urban Areas. *Landscape and Urban Planning*, 77(1–2), 111–130.

Keim, M. E. (2008). Building human resilience: the role of public health preparedness and response as an adaptation to climate change. *American Journal of Preventive Medicine*, 35(5), 508–516.

Kelly, H. (2011). The classical definition of a pandemic is not elusive. *Bulletin of the World Health Organization*, 89, 540–541.

King, R., Proudfoot, L., Smith, B. (1997). *The Mediterranean. Environment and societySociety*. London: Arnold.

Kirkpatrick, I., Ackroyd, S., Walker, R. (2005). The new managerialism and public service professions. Hampshire: Palgrave Macmillan.

Kitchener, M. (1998). Quasi-market transformation: an institutionalist approach to change in UK hospitals. *Public Administration*, 76(1), 37-95.

Kloosterman, R. C., and Musterd S. (2001). The polycentric urban region: towards Towards a research agenda. *Urban Studies*, 38(4), 623–633.

Kourliouros, E. (1997). Planning industrial location in Greater Athens: The interaction between deindustrialization and antiindustrialism during the 1980s. *European Planning Studies*, 5(4), 435–460.

Kovach, C., and Mandell, M. P. (1990). A new public-sector-based model of strategic management for cities. *State & Local Government Review*, 27–36.

Krumholz, N. (1992). Roman impressions: contemporary Contemporary city planning and housing in Rome. *Landscape and Urban Planning*, 22(2–4), 107–114.

Kübler, D., Heinelt, H. (2004). Metropolitan governance, democracy and the dynamics of place. In *Metropolitan Governance in the 21st Century*, 20–40. Routledge.

Lamonica, G.R., Chelli, F.M. (2018). The performance of non-survey techniques for constructing sub-territorial input-output tables. *Papers in Regional Science*, 97(4), 1169-1202.

Leontidou, L. (1990). *The Mediterranean City in Transition*. Cambridge: Cambridge University Press.

Lega, F. (2016). *Management e Leadership dell'Azienda Sanitaria*. Milano: Egea.

Le Grand, J. (1991). Quasi-markets and social policy. *The economic journal*, 101(408), 1256-1267.

Le Grand, J. (1999). Competition, Cooperation, Or Control? Tales From The British National Health Service: In the battle between market competition and central control in Britain's health care system, control won. Will Labour's new version of the market prevail?. *Health affairs*, 18(3), 27-39.

Leontidou, L. (1996). Alternatives to modernism in (southern) urban theory: Exploring in-between spaces. *International Journal of Urban and Regional Research*, 20(2), 178–195.

Longhi, C., Musolesi, A. (2007). European cities in the process of economic integration: Towards structural convergence. *Annals of Regional Science*, 41(2), 333–351.

Losch, B., Fréguin-Gresh, S., White, E. T. (2012). *Structural transformation and rural change revisited: Challenges for Late developing countries in a globalizing world*. The World Bank.

Marchi, M., Ferrara, C., Biasi, R., Salvia, R., Salvati, L. (2018). Agro-forest management and soil degradation in Mediterranean environments: Towards a strategy for sustainable land use in vineyard and Olive olive Croplandcropland. *Sustainability* (Switzerland).

Masini, E., Tomao, A., Barbati, A., Corona, P., Serra, P., Salvati, L. (2019). Urban Growthgrowth, Landland-use Efficiency efficiency and Local local Socioeconomic socioeconomic Contextcontext: A Comparative comparative Analysis analysis of 417 Metropolitan metropolitan Regions regions in Europe. *Environmental Management*, 63(3), 322–337.

Mason, C., Roy, M. J., Carey, G. (2019). Social enterprises in quasi-markets: Exploring the critical knowledge gaps. *Social Enterprise Journal*.

Matvejevic, P. (1998). *Il Mediterraneo e l'Europa*. Milano: Garzanti, 1998. McKibbin, W. J., Sidorenko, A. (2006). Global macroeconomic consequences of pandemic influenza. Sydney, Australia: Lowy Institute for International Policy.

Meijers, E. (2008). Measuring polycentricity and its promises. *European Planning Studies*, 16(9), 1313–1323.

Mitchell, J. (1999). *Crucibles of Hazard: Mega-cities and disasters in transition*. United Nations University Press.

Moretti, V., Salvati, L., Cecchini, M., Zambon, I. (2019). A long-term analysis of demographic processes, socioeconomic "modernization" and forest expansion in a European Countrycountry. *Sustainability* (Switzerland), 112, 388.

Munafò, M., Norero, C., Sabbi, A., Salvati, L. (2010). Urban soil consumption in the growing city: A survey in Rome. *Scottish Geographical Journal*, 126(3), 153–161.

Muñoz, F. (2003) Lock living: Urban sprawl in Mediterranean cities. *Cities*, 20(6), 381–385.

Naudé, W., Santos-Paulino, A. U., and McGillivray, M. (2009). Measuring vulnerability: An overview and introduction. *Oxford Development Studies*, 37(3), 183–191.

Nicita, A. (2004). *Il pendolo delle riforme nei sistemi sanitari europei*, Quaderni CERM, 5/04, April.

Nonis, M., Rossi, E. (2009). L'aggiornamento dei Diagnostic related groups alla versione CMS 24.0 in Italia e il dibattito europeo sul finanziamento dell'attivita' ospedaliera. *Politiche Sanitarie*, 10 (2), April.

Noji, E.K. (1997). *The nature Nature of disasterDisaster: general General characteristics Characteristics and public Public health Health effects Effects*. Oxford University Press, Oxford, United Kingdom: Oxford University Press, 3–20.

Parks, R.B., Oakerson, R.J. (1989). Metropolitan organization and governance: A local public economy approach. *Urban Affairs Quarterly*, 25(1), 18–29.

Paul V., Tonts M. (2005). Containing urban sprawl: trends Trends in land use and spatial planning in the Metropolitan metropolitan Region region of Barcelona. *Journal of Environmental Planning and Management*, 48(1), 7–35.

Peckham, R. (2013). Economies of contagion: financial crisis and pandemic. *Economy and Society*, 42(2), 226–248.

Petretto, A. (2010). *Health care organization models and financial systems*. Working Paper n. 4/2010, Florence: Department of Economic Sciences.

Petretto, A. (2009). *Modelli economici di organizzazione sanitaria e finanziamento. Paper presentato al Convegno "Diritto alla salute tra unità e differenziazione: Modelli di organizzazione sanitaria a confronto"*. Fondazione Cesfin A. Pedrieri, Florence, 20(11), 2009.

Proietti. C., Anav. A., Vitale. M., Fares. S., Fornasier. F., Screpanti. A., Salvati. L., Paoletti. E., Sicard. P., De Marco. A. (2019). A new wetness index to evaluate the soil water availability influence on gross primary production of European forests.2019, *Climate*, 73, 42.

Ranade, W. (1995). The theory and practice of managed competition in the National Health Service. *Public Administration*, 73(2), 241-262.

Richardson, H. W., Chang-Hee C. B. (2004). *Urban sprawl Sprawl in Western Europe and the United States*. London: Ashgate.

Rivolin, U. J., Faludi, A. K. F. (2005). The hidden face of European spatial planning. *European Planning Studies*, 13(2), 195–215.

Rosti, L., Chelli, F. (2009). Self-employment among Italian female graduates. *Education and Training*, 51(7), 526-540.

Rosti, L., Chelli, F. (2012). Higher education in non-standard wage contracts. *Education and Training*, 54(2-3), 142-151.

Salet, W. G., Salet, W. G. M., Thornley, A., Kreukels, A. (2003). *Metropolitan governance and spatial planning: comparative case studies of European city-regions*. Taylor & Francis, London.

Sallustio, L., Pettenella, D., Merlini, P., Romano, R., Salvati, L., Marchetti, M., Corona, P. (2018). Assessing the economic marginality of agricultural lands in Italy to support land use planning. *Land Use Policy*, 76, 526–534.

Salvati, L. (2018). The 'niche' city: A multifactor spatial approach to identify local-scale dimensions of urban complexity. *Ecological Indicators*, 94, 62–73.

Salvati, L., Bajocco, S., Ceccarelli, T., Zitti, M., Perini, L. (2011). Towards a process-based evaluation of land vulnerability to soil degradation in Italy. *Ecological Indicators*, 11(5), 1216–1227.

Salvati, L., Colantoni, A. (2015). Land use dynamics and soil quality in agro-forest systems: A country-scale assessment in Italy. *Journal of Environmental Planning and Management*, 58(1), 175–188.

Salvati, L., Gargiulo Morelli, V. (2014). Unveiling Urban Sprawl in the Mediterranean Region: Towards a Latent Urban Transformation? *International Journal of Urban and Regional Research*, 38(6), 1935–1953.

Salvati, L., Guandalini, A., Carlucci, M., Chelli, F.M. (2017). An empirical assessment of human development through remote sensing: Evidences from Italy. *Ecological Indicators*, 78, 167-172.

Salvati, L., Tombolini, I., Ippolito, A., Carlucci, M. (2018). Land quality and the city: Monitoring urban growth and land take in 76 Southern southern European metropolitan areas. 2018, *Environment and Planning B: Urban Analytics and City Science*, 454(3), 691–712.

Salvati, L., Tombolini, I., Perini, L., Ferrara, A. (2013). Landscape changes and environmental quality: The evolution of land vulnerability and potential resilience to degradation in Italy. *Regional Environmental Change*, 13(6), 1223–1233.

Salvati, L., Zambon, I., Pignatti, G., Colantoni, A., Cividino, S., Perini, L., Pontuale, G., Cecchini, M. (2019). A time-series analysis of climate variability in urban and agricultural sites (Rome, Italy). *Agriculture* (Switzerland), 95, 103.

Salvati, L., Zitti, M., Ceccarelli, T., Perini, L. (2009). Developing a synthetic index of land vulnerability to drought and desertification. *Geographical Research*, 47(3), 280–291.

Schneider, A., Woodcock, C. E. (2008). Compact, dispersed, fragmented, extensive? A comparison of urban growth in twenty-five global cities using remotely sensed data, pattern metrics and census information. *Urban Studies*, 45(3), 659–692.

Soliman, A.M. (2004). Regional planning scenarios in South Lebanon: The challenge of rural-urban interactions in the era of liberation and globalization. *Habitat International*, 28(3), 385–408.

Spina, S., Marrazzo, F., Migliari, M., Stucchi, R., Sforza, A., and Fumagalli, R. (2020). The response of Milan's Emergency Medical System to the COVID-19 outbreak in Italy. *The Lancet*, available on the website https://doi.org/10.1016/S0140-6736(20)30493-1.

Sutton, P. C., Goetz, A. R., Fildes, S., Forster, C., Ghosh, T. (2010). Darkness on the edge of town: mapping urban and peri-urban Australia using nighttime satellite imagery. *The Professional Geographer*, 62(1), 119–133.

Sylves, R. T. (2019). *Disaster policy Policy and politicsPolitics: Emergency Management and Homeland Security*. CQ Press.

Terzi F., Bolen, F. (2009). Urban sprawl measurement of Istanbul. *European Planning Studies*, 17(10), 1559–1570.

Turok, I., Mykhnenko, V. (2007). The trajectories of European cities, 1960–2005. *Cities*, 24(3), 165–182.

Vaughan, E., and Tinker, T. (2009). Effective health risk communication about pandemic influenza for vulnerable populations. *American Journal of Public Health*, 99(S2), S324–S332.

Verikios, G., Sullivan, M., Stojanovski, P., Giesecke, J. A., Woo, G. (2011). The global economic effects of pandemic influenza. Centre of Policy Studies (CoPS).

Jones, P.R., Cullis, J.G. (1996). Decision making in quasi-markets: A pedagogic analysis. *Journal of Health Economics*, 15(2), 187-208.

Waring, S. C., and Brown, B. J. (2005). The threat of communicable diseases following natural disasters: A public health response. *Disaster Management & Response*, 3(2), 41–47.

Weber, C., Petropoulou, C., Hirsch, J. (2005). Urban development in the Athens metropolitan area using remote sensing data with supervised analysis and GIS. *International Journal of Remote Sensing*, 26(4), 785–796.

Wechsler, B., Backoff, R. W. (1986). Policy making and administration in state agencies: Strategic management approaches. *Public Administration Review*, 321–327.

Wilson, E. (2006). Adapting to climate change at the local level: the spatial planning response. *Local Environment*, 11(6), 609–625.

World Health Organization. (2013). *Protecting Health from climate Climate Change: vulnerability and adaptation assessment*. World Health Organization.

Yohe, G. (1990). The cost of not holding back the sea: Toward a national sample of economic vulnerability. *Coastal Management*, 18(4), 403–431.

Zambon, I., Cerdà, A., Cividino, S., Salvati, L. (2019a). The (Evolving) vineyard's age structure in the Valencian community, Spain: A new demographic approach for rural development and landscape analysis.2019a. *Agriculture* (Switzerland), 93, 59.

Zambon, I., Colantoni, A., Carlucci, M., Morrow, N., Sateriano, A., Salvati, L. (2017). Land quality, sustainable development and environmental degradation in agricultural districts: A computational approach based on entropy indexes. *Environmental Impact Assessment Review*, 64, 37–46.

Zambon, I., Colantoni, A., Salvati, L. (2019b). Horizontal vs vertical growth: Understanding latent patterns of urban expansion in large metropolitan regions. *Science of the Total Environment*, 654(1), 778–785, 1.

5

Urban "Crisis" and Management of "Smart" Land: A New Integrated Strategy of Public Governance to Give Value to Local Communities

Sabato Vinci

The recent crisis of Western societies brought challenges and requires innovative approaches to urban planning and design. With this assumption in mind, the contribution of management and public governance sciences to local development should recognize the importance of managing the available land resources according to sustainability criteria, with the aim to create an attractive life context where people might enjoy living and working peacefully (Rosti and Chelli, 2009; Gigliarano and Chelli, 2016; Carlucci et al., 2018; Cecchini et al., 2019), taking into account not only current needs, but also those of future generations. The main focus of the scientific community is how to deal with sustainable development and resilience at local scale (Salvati et al., 2017), offering socioeconomic and managerial alternative models, which could support a turnaround compared to the waste of resources in the past, in order to limit the CO2 emissions and to cope with climatic changes (Francaviglia et al., 2019; Proietti et al., 2019; Salvati et al., 2019).

The central issue regarding any territorial transformation is the reduction of energy consumption, waste management, environment protection, and conservation of natural resources (Scarascia et al., 2006; Monarca et al., 2008a, 2008b; Salvati et al., 2008; Colantoni et al., 2013, 2016; Chelleri et al., 2015; Boubaker et al., 2016; Kosmas et al., 2016; Anifantis et al., 2017; Zambon et al., 2018). These approaches include potential drivers of urban development. In a few years, the "sustainable" term has been replaced by the "smart" notion of urban growth, characterizing e.g. communities, governance, energy, mobility, environment, education, living, and communication (Chelli et al., 2016). The risk involved in such a narrative is an overwhelming rhetoric, with few concrete actions taken by city councils and no improvements for local communities. In order to be "smart", a city has to provide concrete answers of sustainable land and environment management: energy production from renewable sources, improvement of the quality of its own environment, pollution prevention (e.g. De Marco et al., 2019), making services accessible to the citizens (Carlucci et al., 2018), either through the elimination of architectural barriers, and through open data (Monarca et al., 2011a, 2011b, 2012; Febbi et al., 2015; Marucci et al., 2018; Zambon et al., 2018).

A smart city consists of an integrated model of public governance capable to provide alternative mobility systems, and to promote the use of bicycle, public transports, car sharing, and pedestrian mobility (Bifulco et al., 2016). Additionally, a smart city provides parks and green spaces, with the purpose of enhancing the quality of public spaces and urban liveliness (Meijer, 2016). Generally, smart projects for smart cities provide appropriate interventions and investments in physical and digital communication infrastructure that affect urban mobility and information flows (Cecchini et al., 2015; Zambon et al., 2016; Colantoni et al., 2018).

Above all, there is the idea that Information and Communication Technologies (ITC) might increase the efficiency of different urban systems thanks to a remote management issue and the interface with the users, which provokes savings upon money and time and generate other development sectors in the economy. Anyway, there is always one single and recurring question that concerns technological innovation, wondering how much technology we actually need in order to satisfy the users' needs, if the investments are commensurate with the purpose, and whether there are reliable benefits for the collective well-being (Chelli and Rosti, 2002; Rosti and Chelli, 2012; Ciommi et al., 2017; Lamonica and Chelli, 2018). A sustainable urban management should consider the necessity of its community to develop cooperative strategies upon a shared vision of the future (centered community planning). Moreover, it has to be able to activate virtuous processes of economic development and social innovations (e.g. Chelli et al., 2009), which might generate improvements to the quality of life of resident populations (smart urban planning).

5.1 From "Smart City" to "Smart Land"

The well-known model of smart cities implies a specific risk of generalization without any correlation to local identities, exporting *tout court* practices of local development that are not focused on specific citizen's needs of that particular socioeconomic context. Cities are different as they have different development paths; they pursue different goals and generate distinctive relationships within their surrounding territory (Ciommi et al., 2018). For instance, it can be recalled the case of the Adriatic coast in Italy, where cities follow one another without any spatial break; the case of roads linking urban centers, which become axes of concentration of economic activities; the case of mountain communities that share large territories and natural landscapes.

It is possible to move from the concept of "smart city" to the notion of "smart land" (Bonomi, 2014), since the boundaries between urban and rural areas are becoming more and more flexible, leading to new ways of rethinking peri-urban land and new agricultural functions against urbanization. In these regards, it is essential to remark the context of urban planning at different levels: from wide areas ("smart land") to the "local districts" which means reduced portion of urban or rural territories, where strategic projects can be developed. The concept of local district has been used at various extents and with different meanings: industrial and productive districts (Marshall, 1920; Becattini, 1989, 2000; Bagnasco, 1977); regional districts and urban districts (Collier, 1972; Brusco, 1989; Feiock, 2007); and agro-forest and rural districts (Salvati et al., 2011; Recanatesi et al., 2016). Generally, these districts concern just some portions of the territory, which present similar features and shared economic interests. Piore and Sabel (1984) and Porter (1989) reinterpret the "district," more generally, as a "cluster" of economic activities.

From the economic point of view, Marshall was likely the first scholar to identify the advantages connected with the industrial districts, studying the companies of Yorkshire and Sheffield, specialized in wool and metal processing: "When an industry has thus chosen a locality for itself, it is likely to stay there long: so great are the advantages which people following the same skilled trade get from near neighbourhood to one another. The mysteries of the trade become no mysteries; but are as it were in the air, and children learn many of them unconsciously. Good work is rightly appreciated; inventions and improvements in machinery, in processes, and the general organization of the business have their merits promptly discussed: if one man starts a new idea, it is taken up by others and combined with suggestions of their own; and thus it becomes the source of further new ideas. And presently subsidiary trades grow up in the neighbourhood, supplying it with implements and materials, organizing its traffic, and in many ways conducing to the economy of its material" (Marshall, 1920).

Based on Marshall's studies, the international economic literature, starting with Porter (Porter 1990; Enright 1992, 1993; Ketels, 2006), highlighted a certain convergence between concepts of "district" and "cluster". On the one hand, both concepts would indicate economic activities concentrated on the same geographic area, which are able to provide to "clusterized" or "districtized" enterprises greater advantages than those single enterprises not included in a district or cluster are able to achieve. On the other hand, it was noted that the notion of cluster is broader, being inclusive of the notion of 'district'. In particular, according to Enright (1996), the fundamental peculiarity of the district, compared to the cluster, is represented by the role played by local communities in relation with the local enterprises. In practice, the reference is to the ability of the local population to connect with a productive apparatus, providing a fundamental support identifiable in the values and cultural system of the local population (Becattini and Coltorti 2004; Becattini and Bellandi, 2002). These elements are able to pervade the local industrial system and to provide it the engine for economic development (Ciommi et al., 2019).

The meaning of ecological and cultural districts is likely more trasversal. In regard with these last ones, there have been interesting experiences in Southern Europe aimed at safeguarding together cultural and environmental values, receiving subsidies directed to promotion of natural, archeological, and artistic assets. A "district" could be considered a territory diversified and complex enough to share identity, landscape, cultural and social values, besides the economic use. Within a district, many functions may converge that could involve centers for innovation and research, services for leisure, craftsmanship, cultural attractions and tourism, according to a common vision that identifies a recognizable image. In fact, it is possible to talk about a district when its internal relationships and social knots are defined by the intersection of more than one theme. Moreover, the use of new technologies (energy-saving technologies, interconnected terminals, integrated knowledge, waste recovery, time optimization, and increase of services' efficiency) should be coupled with the social innovation that technology can bring in a landscape perspective (Cecchini et al., 2010a, 2010b; Di Giacinto et al., 2012; Zambon et al., 2018).

"Social innovations" can be considered as the new solutions (products, services, models, markets, and processes), which, on the one hand, satisfy a social need in a more efficient and effective way, in comparison with the existing ones. On the other hand, they lead to new (or improved) functionalities, activities, and/or relationships. To sum up, social innovations improve the quality of the whole community and its own capacity of acting within its borders. Social innovation involves changes in culture and practice in order to cooperate for a more sustainable society: from new working models to new projects and solutions that address local needs, using available resources and knowledge. Local governance actions enable the connections of the material infrastructures of the city with its human, intellectual, and social capital. Thanks to the adoption of new communication technologies, efficient mobility systems, environmental safeguard and energy saving, it is possible to satisfy the needs of citizens, enterprises, and institutions.

5.2 From "Resilient Districts" to "Sustainable City"

The concept of resilient city and district is similar, but different from the notion of smart city and smart district. The general framework of smart city and resilient city presents similar goals to achieve, when they are considered in the following dimensions:

- Economy and society: The economic and social system should provide all the people to live in peace and act collectively.
- Infrastructure and environment: The system of infrastructures should guarantee an efficient flow of people, goods, and knowledge; environmental quality should be also considered.
- Health and well-being: anyone who is working either living in the city should have direct access to first-need goods, services, and job opportunities.
- Leadership and strategy: all the citizens might have the chance to participate in the decision process through inclusive planning and stakeholder's empowerment.

In this regard, a part of the management literature has already emphasized the concept of "collaborative public governance" or "governance wave" (Agranoff et al., 2003; Bryson et al., 2006; McGuire, 2006; Provan et al., 2008; Emerson et al., 2012). This phenomenon arise from a progressive weakening of the centralized State model in the "liquid" modern society and it is embodied in the public institutions' ability to directly involve "non-public" stakeholders in a collective decision-making process (Dawkins, 2015; Zangrandi, 2019).

While the smart city aims at the elimination of "repetitions" which represent a high community cost, the resilient city points at redundancy and "diversification" in order to create a portfolio of different alternatives (e.g. Di Feliciantonio et al., 2018; Carlucci et al., 2019). Being more careful on sharing common objectives than achieving individual results, the resilient districts are focused on local cooperation logic more than strategies based on competition. The term resilience comes from the metallurgy sector, defining the capacity of a metal to absorb the energy of elastic deformation. In other words, it means the capacity of a metal to get back to a new condition of equilibrium after enduring external strains. In psychology, it explains the ability to react to traumas in a positive way so that human brain can regenerate itself. In ecology, this notion refers to the capacity of a living element to recover itself after a shock, or the ability of an ecological system to react to disturbance regimes, which have caused an alteration of its balanced conditions (Ciaccia et al., 2019).

Resilience is a property of complex systems reacting to stresses. It happens through the activation of responding strategies and adaptation, in order to restore the operating mechanisms. Resilient systems react to external stresses renewing themselves but retaining functions typical of the system (Holling and Gunderson, 2002). Resilience involves restoration of functions through mutation and adaptation. In relation to the city, resilience could be defined as the art of adapting to changes, transforming uncertainties into opportunities, and risks into innovation. The Latin root "resilire" (literally standing for "to bounce back") suggests a turnaround following a bounce, or to resist impacts without bending. During these years' crisis, the city or community resilience can be described as the ability of a system to survive, adapt, and develop while enduring negative scenarios. So, it explains how good a city or a community has to be to seize ascents (opportunities) after negative events. Urban resilience, in slightly different terms than "smart city", has to withstand climate change, but also human pressure and urban transformations (Biasi et al., 2019; Francaviglia et al., 2019; Proietti et al., 2019; Salvati et al., 2019).

"A Resilient City is one that has developed capacities to help absorb future shocks and stresses to its social, economic, and technical systems and infrastructures so as to still be able to maintain essentially the same functions, structures, systems, and identity" (ResilientCity.org).

Within urban systems, resilience is a meaningful concept to explain those path-dependent processes which are based on peculiarities that enable the territory itself to react to external shocks throughout autonomous rearrangement (Hopkins, 2008). Boschma and Martin (2007), and other authors referring to the "Evolutionary Economic Geography" school, investigated the economic, social, cultural, and institutional features which determine the capacity of a territory to reply to external incentives (e.g. economic change), creating new development paths. Particularly, a resilient system should be redundant, diversified, efficient, independent, adaptable, collaborative, providing control feedback as well (Colucci, 2012; Luciani et al, 2018). Diversity and creative redundancy of the functions are central factors to ensure resilience (Low et al., 2003; Barthel et al., 2013; Cording et al., 2013). A resilient socioeconomic system should also be compatible with different values, and satisfy multiple goals, ensuring the operation of subsystems in a non-hierarchical manner, in order to minimize the risk factors and to guarantee alternatives.

Urban planning and governance, in terms of resilience, considers the production capacity of the territory and the equilibrium condition of rural systems, in relation with urban consumption: a sustainable management approach, developing food supply chains, energy self-sufficiency, and waste treatment efficiency (Colantoni et al., 2013, 2016; Febbi et al., 2015; Boubaker et al., 2016; Anifantis et al., 2017). In addition, resilient planning and governance should assure a thorough reduction of the risks and the strengthening of the capacity to resist (Morrow et al., 2018; Cecchini et al., 2019). Concerning the food security issue, the resilient district focuses on bringing back a certain balance within urban– rural systems, supporting peri-urban agriculture and the implementation of innovative techniques to improve agricultural production capacity at the local scale (Cecchini et al., 2013; Marucci et al., 2012, 2013; Moscetti et al., 2015a, 2015b). Then, resilience of a given territory generates interests for periurban countryside and rural landscape (Barbati et al., 2013; Kosmas et al., 2016; Serra et al., 2018).

The preservation and the enhancement of rural landscapes throughout the promotion of fresh local products for urban population and the development of a stronger connection between the environmental dimension and local communities are the real challenge for the European countries that want to achieve a sustainable development (Daniels, 2016). Considering this perspective, the present crisis might be seen as a great opportunity for improving the intrinsic relationship between economy, environment, and production choices, as suggested by Florida (2011), but also as a strong policy challenge for public institutions.

5.3 Sustainable Landscape Management as Social and Economic Opportunity

Landscapes, considering their wide meaning, have a very important economic and industrial value, and, of course, a cultural and environmental value. In fact, they play a role of a driver for change of local territories and the related production systems, and they can count on particular values that cannot be reproduced with international competition. So, the landscapes are able to attract cultural tourism, generating large benefits for commercial and production activities and the quality of life of the citizens. Throughout the recognition and the evaluation of a landscape identity within its territory, we can promote governance choices and marketing strategies, to clarify the relationship between landscape and product, and start offering some services, which might attract quality consumption and tourism. It means to overlay and combine several layers, such as the environmental protection with cultural attractions; slow mobility networks with services, production activities, wine and food goods, and cultural events (Marucci et al., 2012, 2013; Cecchini et al., 2013; Moscetti et al., 2015a, 2015b).

5.3.1 Collective Identities, Strategic Visions, and Synergic Actions

Local landscape can be considered the engine for innovation and social cooperation for the creation of both smart land and resilient districts. It refers to the idea of sharing visions and strategies of local development throughout sustainable actions, which will take their strength directly from social cohesion, proposals for making creative cities coming from the bottom. The landscape is the perfect ground for sharing collective identities, strategic visions, and synergic actions. A landscape could be seen as the big context, with the same meaning used for the social networks, on which it is possible to introduce strategies to gain back the public space and to recognize a collective identity, to build a territory socially responsible, and to introduce sustainable transformations (Zambon et al., 2017).

Paying attention to the landscape means promoting policies shared and spread, aimed at improving competitiveness and attraction, social cohesion, spread of knowledge, creative growth, accessibility to the environment (natural, historical, and architectural), and quality of life. Many cities seem to exploit the potentials of their territories. They perceive a different identity and end up developing nonlinear processes of transformation, which will reveal not perfectly fitting the local economies. The identity of a place, a neighborhood, or a city, can be defined (i) through architectural features related to the spatial structure of the urban fabric, and (ii) through social practices, narratives, human activities, which are, by definition, so many that might suggest different points of view. Smart land and resilient districts are definitely territorial settings, which share a common vision of the future.

The involvement of local people in the formation of districts that resist crisis and share sustainable development strategies can be done through the exercise of "territorial foresight". It deals with a series of practices that go far beyond from public participation, which is typical of regional planning. It is not a question of predicting the future, neither to produce more detailed assessment techniques, or more advanced interpretative models. The scope of the foresight practice is to imagine the evolution of a given landscape, including the available resources and the natural potentialities of a territory, thanks to technological and cultural changes. The foresight aims at the creation of perspectives for local development through a systematic and inclusive process, which might strategically guide planning choices, look for proper economic means, and favor joined actions. It tries to "bites the time", developing strategic visions and gathering the stakeholders, which are the real protagonists of those changes. Once they met, they could work for a new sharing process, which may eventually bring to information surveys and the creation of long- or mid-run perspectives that can be addressed for present decisions. Strategic foresight visions can be considered as the guiding lines that can be proved as catalyst for ideas, resources, synergic and collective engagement (Di Feliciantonio et al., 2018; Morrow et al., 2018; Cecchini et al., 2019; Carlucci et al., 2019).

5.4 Conclusions

Landscape identity is a key element for the economic and cultural enhancement of the territory, through governance processes consistent with the environment and traditions of local communities. The landscape can achieve shared interests, empathetic membership toward common values, facilitating diversification, and preserving system's redundancy related to regional resilience. Landscape projects should be able to create a strategic vision and the scenario for an integrated territorial development, able to promote the potential of local development, strengthening the adaptive skills of the places to the evolutionary dynamics. Landscapes are able to attract cultural tourism, generating large benefits for businesses and industrial activities, as well as for citizens' quality of life. Throughout recognition and evaluation of a landscape identity within its territory, we can promote governance choices and marketing strategies to clarify the relationship between landscape and products, and start offering some services attracting quality consumption and tourism.

Therefore, a strategic management of the landscape, in the most positive of its meanings, suggests an interventation mode that reduce the probability for a 'shock' of socio-environmental ecosystems, instead preferring a development plan coherent with environmental balances and value systems that characterize the local identity.

References

Adger, W. N. (2000). Social and ecological resilience: Are they related? *Progress in Human Geography*, 24, 347.

Agranoff, R., McGuire, M. (2003). *Collaborative Public Management: New Strategies for Local Governments*. Georgetown University Press.

Anifantis, A. S., Colantoni, A., Pascuzzi, S. (2017). Thermal energy assessment of a small scale photovoltaic, hydrogen and geothermal stand-alone system for greenhouse heating. *Renewable Energy*, 103, 115–127.

Bagnasco, A. (1977). *Tre Italie: la problematica territoriale dello sviluppo italiano*. Il Mulino, Bologna.

Barbati, A., Corona, P., Salvati, L., Gasparella, L. (2013). Natural forest expansion into suburban countryside: Gained ground for a green infrastructure? *Urban Forestry and Urban Greening*, 12(1), 36–43.

Barthel, S., Isendahl, C. (2013). Urban gardens, agriculture, and water management: Sources of resilience for long-term food security in cities. *Ecological Economics*, 86, 224–234.

Becattini, G. (1989). Riflessioni sul distretto industriale marshalliano come concetto socio-economico. *Stato e mercato*, 111–128.

Becattini, G. (2000). *Il distretto industriale: un nuovo modo di interpretare il cambiamento economico*. Rosenberg & Sellier.

Becattini, G., Bellandi, M. (2002). Forti Pigmei e deboli Vatussi. Considerazioni sull'industria italiana. *Economia italiana*, 3, 587-618.

Becattini, G., Coltorti, F. (2004). Aree di grande impresa ed aree distrettuali nello sviluppo post-bellico dell'Italia: un'esplorazione preliminare. *Rivista Italiana degli Economisti*, 9(1), 61-102.

Biasi R., Brunori E., Ferrara C., Salvati L. (2019). Assessing impacts of climate change on phenology and quality traits of *Vitis vinifera* L.: The contribution of local knowledge. *Plants*, 85, 121.

Bifulco, F., Tregua, M., Amitrano, C. C., D'Auria, A. (2016). ICT and sustainability in smart cities management. *International Journal of Public Sector Management*, 29(2), 132–147.

Bonomi, A., Masiero R. (2014). *Dalla smart city alla smart land*. Venezia: Marsilio Editori.

Boubaker, K., Colantoni, A., Allegrini, E., Longo, L., Di Giacinto, S., Monarca, D., Cecchini, M. (2014). A model for musculoskeletal disorder-related fatigue in upper limb manipulation during industrial vegetables sorting. *International Journal of Industrial Ergonomics*, 44(4), 601–605.

Boubaker, K., Colantoni, A., Marucci, A., Longo, L., Gambella, F., Cividino, S., Cecchini, M. (2016). Perspective and potential of CO2: A focus on potentials for renewable energy conversion in the Mediterranean basin. *Renewable Energy*, 90, 248–256.

Boyle, R., Mohamed, R. (2007). State growth management, smart growth and urban containment: A review of the US and a study of the heartland. *Journal of Environmental Planning and Management*, 50(5), 677–697.

Briassoulis, H. (2005). *Policy Integration for Complex Environmental Problems*. Aldershot: Ashgate.

Bristow, G., Wells, P. (2005). Innovative discourse for sustainable local development: A critical analysis of eco-industrialism. *International Journal of Innovation and Sustainable Development*, 1, 168–179.

Brusco, S. (1989). *Piccole imprese e distretti industriali: una raccolta di saggi*. Rosenberg & Sellier.

Bryson, J. M., Crosby, B. C., Stone, M. M. (2006). The design and implementation of cross-sector collaborations: Propositions from the literature. *Public Administration Review*, 66, 44–55.

Carlucci, M., Chelli, F.M., Salvati, L. (2018). Toward a new cycle: Short-term population dynamics, gentrification, and re-urbanization of Milan (Italy). *Sustainability (Switzerland)* 10(9), 3014.

Carlucci, M., Zambon, I., Salvati, L. (2019). Diversification in urban functions as a measure of metropolitan complexity. *Environment and Planning B: Urban Analytics and City Science*, 1–17.

Carta M. (2007). *Creative City. Dinamics, Innovations, Actions*. Trento: List.

Carta, M. (2011). Città creativa 3.0. Rigenerazione urbana e politiche di val-orizzazione delle armature culturali, in M. Cammelli, P.A. Valentino (a cura di), *Citymorphosis. Politiche culturali per città che cambiano*, 213–22. Firenze: Giunti.

Cassa Depositi e Prestiti (2013). *Smart city. Progetti di sviluppo e strumenti di finanziamento*.

Cecchini, M., Cividino, S., Turco, R., Salvati, L. (2019). Population age structure, complex socio-demographic systems and resilience potential: A spatio-temporal, evenness-based approach. *Sustainability* (Switzerland), 117, 20–50.

Cecchini, M., Colantoni, A., Massantini, R., Monarca, D. (2010a). The risk of musculoskeletal disorders for workers due to repetitive movements during tomato harvesting. *Journal of Agricultural Safety and Health*, 16(2), 87–98.

Cecchini, M., Colantoni, A., Massantini, R., Monarca, D. (2010b). Estimation of the risks of thermal stress due to the microclimate for manual fruit and vegetable harvesters in central Italy. *Journal of Agricultural Safety and Health*, 16(3), 141–159.

Cecchini, M., Cossio, F., Marucci, A., Monarca, D., Colantoni, A., Petrelli, M., Allegrini, E. (2013). Survey on the status of enforcement of European directives on health and safety at work in some Italian farms. *Journal of Food, Agriculture, and Environment* 11, 595–600.

Cecchini, M., Monarca, D., Colantoni, A., Cossio, F., Moscetti, R., Massantini, R., Bedini, R. (2015). "Smart Bio Wine", a Project for Consumer Information, Business Competitiveness and Environmental Sustainability. *Chemical Engineering*, 44.

Chelleri, L., Schuetze, T., Salvati, L. (2015). Integrating resilience with urban sustainability in neglected neighborhoods: Challenges and opportunities of transitioning to decentralized water management in Mexico city. *Habitat International*, 48, 122–130.

Chelli, F., Rosti, L. (2002). Age and gender differences in Italian workers' mobility. *International Journal of Manpower*, 23(4), 313-325.

Chelli, F., Gigliarano, C., Mattioli, E. (2009). The impact of inflation on heterogeneous groups of households: An application to Italy. *Economics Bulletin*, 29(2), 1276-1295.

Chelli, F.M., Ciommi, M., Emili, A., Gigliarano, C., Taralli, S. (2016). Assessing the Equitable and Sustainable Well-Being of the Italian Provinces. *International Journal of Uncertainty, Fuzziness and Knowlege-Based Systems*, 24, 39-62.

Ciaccia, C., La Torre, A., Ferlito, F., Testani, E., Battaglia, V., Salvati, L., Roccuzzo, G. (2019). Agroecological practices and agrobiodiversity:

A case study on organic orange in southern Italy. *Agronomy*, 92, 85.
Ciommi, M., Gigliarano, C., Emili, A., Taralli, S., Chelli, F.M. (2017). A new class of composite indicators for measuring well-being at the local level: An application to the Equitable and Sustainable Well-being (BES) of the Italian Provinces. *Ecological Indicators*, 76, 281-296.
Ciommi, M., Chelli, F.M., Carlucci, M., Salvati, L. (2018). Urban growth and demographic dynamics in southern Europe: Toward a new statistical approach to regional science. *Sustainability (Switzerland)*, 10(8), 2765.
Ciommi, M., Chelli, F.M., Salvati, L. (2019). Integrating parametric and non-parametric multivariate analysis of urban growth and commuting patterns in a European metropolitan area. *Quality and Quantity*, 53(2), 957-979.
Colantoni, A., Allegrini, E., Boubaker, K., Longo, L., Di Giacinto, S., Biondi, P. (2013). New insights for renewable energy hybrid photovoltaic/wind installations in Tunisia through a mathematical model. *Energy Conversion and Management*, 75, 398–401.
Colantoni, A., Delfanti, L., Recanatesi, F., Tolli, M., Lord, R. (2016). Land use planning for utilizing biomass residues in Tuscia Romana (central Italy): Preliminary results of a multicriteria analysis to create an agroenergy district. *Land Use Policy*, 50, 125–133.
Colantoni, A., Marucci, A., Monarca, D., Pagniello, B., Cecchini, M., Bedini, R. (2012). The risk of musculoskeletal disorders due to repetitive movements of upper limbs for workers employed to vegetable grafting. *Journal of Food, Agriculture, and Environment,* 10, 14–18.
Colantoni, A., Monarca, D., Laurendi, V., Villarini, M., Gambella, F., Cecchini, M. (2018). Smart machines, remote sensing, precision farming, processes, mechatronic, materials and policies for safety and health aspects. *Agriculture*, 8(4), 47.
Colding, J., Barthel, S. (2013). The potential of 'Urban Green Commons' in the resilience building of cities. *Ecological Economics*, 86, 156–166.
Collier, R. W. (1972). The evolution of regional districts. *BC Studies: The British Columbian Quarterly*, 15, 29–39.
Colucci A. (2012). *Le città resilienti: approcci e strategie*. Polo Jean Monnet, Università di Pavia.
Daniels, T., Lapping, M. (2016). Land preservation: An essential ingredient in smart growth. In: *Growth Management and Public Land Acquisition*, Routledge, 23–48.
Dawkins, C. (2015). Agonistic pluralism and stakeholder engagement. *Business Ethics Quarterly*, 25(1), 1–28.

De Marco, A., Proietti, C., Anav, A., Ciancarella, L., D'Elia, I., Fares, S., Fornasier, M. F., Fusaro, L., Gualtieri, M., Manes, F., Marchetto, A., Mircea, M., Paoletti, E., Piersanti, A., Rogora, M., Salvati, L., Salvatori, E., Screpanti, A., Vialetto, G., Vitale, M., Leonardi, C. Impacts of air pollution on human and ecosystem health, and implications for the National Emission Ceilings Directive: Insights from Italy 2019, *Environment International*, 125, 320–333.

Dematteis, G. (2005). *Territorialità, sviluppo locale, sostenibilità: il modello SLOT*. Milano: Franco Angeli.

Di Feliciantonio, C., Salvati, L., Sarantakou, E., Rontos, K. (2018). Class diversification, economic growth and urban sprawl: Evidences from a pre-crisis European city. *Quality and Quantity*, 52(4), 1501–1522.

Di Giacinto, S., Colantoni, A., Cecchini, M., Monarca, D., Moscetti, R., Massantini, R. (2012). Dairy production in restricted environment and safety for the workers. *Industrie Alimentari*, 530, 5–12.

Dominici, G. (2012). *Smart cities e smart communities: l'innovazione che cresce dal basso*. Forum PA. available on the website https://www.forumpa.it/citta-territori/smart-cities-e-communities-linnovazione-nas ce-dal-basso/.

Emerson, K., Nabatchi, T., Balogh, S. (2012). An integrative framework for collaborative governance. *Journal of Public Administration Research and Theory*, 22(1), 1–29.

Enright, M.J. (1992). Why clusters are the way to win the game? *World Link*, 5, 24-25.

Enright, M. J. (1993). *The Geographic Scope of Competitive Advantage*, in E. Dirven, J. Groenewegen, S. van Hoof (eds.), *Stuck in the Region? Changing Scales of Regional Identity*, Utrecht, 1993 (ń*Netherlands Geographical Studies*ż, 155), pp. 87-102.

Enright M. J. (1996). *Regional Clusters and Economic Development: A Research Agenda*, in U. H. Staber, N. V. Shaefer, B. Sharma (eds.), *Business Networks: Prospects for Regional Development*, Berlin-New York, de Gruyter, 1996.

Febbi, P., Menesatti, P., Costa, C., Pari, L., Cecchini, M. (2015). Automated determination of poplar chip size distribution based on combined image and multivariate analyses. *Biomass and Bioenergy*, 73, 1–10.

Feiock, R. C. (2007). Rational choice and regional governance. *Journal of Urban Affairs*, 29(1), 47–63.

Florida, R. (2011). *The Great Reset: How New Ways of Living and Working Drive Postcrash Prosperity*. Florida: Random House Canada.

Folke, C., Colding, J., Berkers F. (2003). *Navigation Social-ecological Systems*. Cambridge: Cambridge University Press.

Francaviglia, R., Di Bene, C., Farina, R., Salvati, L., Vicente-Vicente, J. L. (2019). Assessing "4 per 1000" soil organic carbon storage rates under Mediterranean climate: A comprehensive data analysis. *Mitigation and Adaptation Strategies for Global Change*, 24(5), 795–818.

Fusero, P. (2010). Smart cities: Intelligent territories and infrastructure for the future. In: *Hiper Adriatica*. Barcellona: Actar-D List.

Fusero, P., Massimiano, L. (2012). *Smart Cities, XV conferenza nazionale società Italiana Urbanisti. "L'urbanistica che cambia: rischi e valori"*. Pescara.

Gigliarano, C., Chelli, F.M. (2016). Measuring inter-temporal intragenerational mobility: an application to the Italian labour market. *Quality and Quantity*, 50(1), 89-102.

Gunderson, L.H., Holling C.S. (2002). *Panarchy: Understanding Transformations in Human and Natural Systems*. Washington: Island Press.

Goodman, E., Bamford, J., Saynor, P. (2016). *Small firms and industrial districts in Italy*. Routledge, London.

Holling, C. S., Gunderson L. H. (2002). Resilience and adaptive cycles. In Gunderson L. H. and Holling C. S. (Eds.), *Panarchy: Understanding* Press.

Hopkins, R. (2008). *The Transition Handbook. From Oil Dependency to Local Resilience*. Chelsea: Green Books Devon Ldt.

Hudson, R. (2009). Resilient regions in an uncertain world: Wishful thinking or a practical reality. *Cambridge Journal of Regions, Economy and Society*, 3, 11–25.

Ketels, C. (2003). The Development of the cluster concept–present experiences and further developments. In *NRW Conference on Clusters, Duisberg, Germany* (Vol. 5).

Kosmas, C., Karamesouti, M., Kounalaki, K., Detsis, V., Vassiliou, P., Salvati, L. (2016). Land degradation and long-term changes in agro-pastoral systems: An empirical analysis of ecological resilience in Asteroussia-Crete (Greece). *Catena*, 147, 196–204.

Lamonica, G.R., Chelli, F.M. (2018). The performance of non-survey techniques for constructing sub-territorial input-output tables. *Papers in Regional Science*, 97(4), 1169-1202.

Low, B., Ostrom, E., Simmon, C., Wilson, J. (2003). Redundancy and diversity: Do they influence optimal management? In Folke C., Colding J. and Berkers F. (Eds.), *Navigation Social-ecological Systems*. Cambridge: Cambridge University Press.

Luciani, A., Del Curto, D. (2018). Towards a resilient perspective in building conservation. *Journal of Cultural Heritage Management and Sustainable Development*, 8(3), 309–320.

Magnaghi, A. (1998). *Il territorio degli abitanti. Società locali e autosostenibilità*. Milano: Masson.

Marshall, A. (1920). *Principles of Economics*. London: Mcmillan.

Marucci, A., Monarca, D., Cecchini, M., Colantoni, A., Biondi, P., Cappuccini, A. (2013). The heat stress for workers employed in laying hens' houses. *Journal of Food, Agriculture, and Environment* 11, 20–4.

Marucci, A., Pagniello, B., Monarca, D., Cecchini, M., Colantoni, A., Biondi, P. (2012). Heat stress suffered by workers employed in vegetable grafting in greenhouses. *Journal of Food, Agriculture, and Environment,* 10, 1117–21.

Marucci, A., Zambon, I., Colantoni, A., Monarca, D. (2018). A combination of agricultural and energy purposes: Evaluation of a prototype of photovoltaic greenhouse tunnel. *Renewable and Sustainable Energy Reviews*, 82, 1178–1186.

McGuire, M. (2006). Collaborative public management: Assessing what we know and how we know it. *Public Administration Review*, 66, 33–43.

Meijer, A., Bolívar, M. P. R. (2016). Governing the smart city: A review of the literature on smart urban governance. *International Review of Administrative Sciences*, 82(2), 392–408.

Monarca, D., Cecchini, M., Colantoni, A. (2011a). Plant for the production of chips and pellet: Technical and economic aspects of a case study in the central Italy. In *International Conference on Computational Science and Its Applications*. Springer, Berlin, Heidelberg, 296–306.

Monarca, D., Cecchini, M., Colantoni, A., Marucci, A. (2011b). Feasibility of the electric energy production through gasification processes of biomass: Technical and economic aspects. In *International Conference on Computational Science and Its Applications*. Springer, Berlin, Heidelberg, 307–315.

Monarca, D., Cecchini, M., Guerrieri, M., Colantoni, A. (2008b). Conventional and alternative use of biomasses derived by hazelnut cultivation and processing. In *VII International Congress on Hazelnut*, 845, 627–663.

Monarca, D., Cecchini, M., Guerrieri, M., Santi, M., Bedini, R., Colantoni, A. (2008a). Safety and health of workers: Exposure to dust, noise and vibrations. In *VII International Congress on Hazelnut*, 845, 437–442.

Monarca, D., Colantoni, A., Cecchini, M., Longo, L., Vecchione, L., Carlini, M., Manzo, A. (2012). Energy characterization and gasification of biomass derived by hazelnut cultivation: Analysis of produced syngas by gas chromatography. *Mathematical Problems in Engineering*.

Morrow N., Salvati L., Colantoni A., Mock N. Rooting the future; On-farm trees' contribution to household energy security and asset creation as a resilient development pathway-evidence from a 20-year panel in rural Ethiopia". *Sustainability* (Switzerland), 1012, 4716.

Moscetti, R., Radicetti, E., Monarca, D., Cecchini, M., Massantini, R. (2015a). Near infrared spectroscopy is suitable for the classification of hazelnuts according to protected designation of origin. *Journal of the Science of Food and Agriculture*, 95(13), 2619–2625.

Moscetti, R., Saeys, W., Keresztes, J. C., Goodarzi, M., Cecchini, M., Danilo, M., Massantini, R. (2015b). Hazelnut quality sorting using high dynamic range short-wave infrared hyperspectral imaging. *Food and Bioprocess Technology*, 8(7), 1593–1604.

Pierre, J. (2011). *The Politics of Urban Governance*. Macmillan International Higher Education.

Piore, M. J., Sabel, C. F. (1984). *The Second Industrial Divide: Possibilities for Prosperity*. New York: Basic Books.

Porter, M. E. (1989). *The Competitive Advantage of Nations and their Firms*. New York: Free Press.

Proietti, C., Anav, A., Vitale, M., Fares, S., Fornasier, F., Screpanti, A., Salvati, L., Paoletti, E., Sicard, P., De Marco, A. (2019). A new wetness index to evaluate the soil water availability influence on gross primary production of European forests. *Climate*, 73, 42.

Provan, K. G., Kenis, P. (2008). Modes of network governance: Structure, management, and effectiveness. *Journal of Public Administration Research and Theory*, 18(2), 229–252.

Purcell, M. (2009). Resisting neoliberalisation: Communicative planning or counterhegemonic movements? *Planning Theory*, 8, 140–165.

Recanatesi, F., Clemente, M., Grigoriadis, E., Ranalli, F., Zitti, M., Salvati, L. (2016). A fifty-year sustainability assessment of Italian agro-forest districts. *Sustainability*, 8(1), 32.

Rizzi, F. (2013). *Smart city, smart commnuity, smart specialization per il management della sostenibilità*. Milano: Franco Angeli.

Rosti, L., Chelli, F. (2009). Self-employment among Italian female graduates. *Education and Training*, 51(7), 526-540.

Rosti, L., Chelli, F. (2012). Higher education in non-standard wage contracts. *Education and Training*, 54(2-3), 142-151.

Rodrìguez Bolìvar, M. P. (2018). Governance models and outcomes to foster public value creation in smart cities. *Scienze Regionali*, 17(1), 57–80.

Salvati, L., Zambon, I., Pignatti, G., Colantoni, A., Cividino, S., Perini, L., Pontuale, G., Cecchini, M. (2019). A time-series analysis of climate variability in urban and agricultural sites (Rome, Italy). *Agriculture* (Switzerland), 95, 103.

Salvati, L., Carlucci, M. (2011). The economic and environmental performances of rural districts in Italy: Are competitiveness and sustainability compatible targets? *Ecological Economics*, 70(12), 2446–2453.

Salvati, L., Petitta, M., Ceccarelli, T., Perini, L., Di Battista, F., Scarascia, M. E. V. (2008). Italy's renewable water resources as estimated on the basis of the monthly water balance. *Irrigation and Drainage: The Journal of the International Commission on Irrigation and Drainage*, 57(5), 507–515.

Scarascia, M. V., Battista, F. D., Salvati, L. (2006). Water resources in Italy: Availability and agricultural uses. *Irrigation and Drainage*, 55(2), 115–127.

Serra, P., Saurí, D., Salvati, L. (2018). Peri-urban agriculture in Barcelona: Outlining landscape dynamics vis-à-vis socio-environmental functions. *Landscape Research*, 435(5), 613–631.

Sforzi, F. (2008). Il distretto industriale: da Marshall a Becattini. In *Il pensiero economico italiano*, XVI, 2, 71–80.

Simeonova, V., Van der Valk, A. (2009). The need for a communicative approach to improve environmental policy integration in urban land use planning. *Journal of Planning Literature*, 23(3), 241–261.

Simmie, J., Martin, R. (2010). The economic resilience of regions: Towards an evolutionary approach. *Cambridge Journal of Regions, Economy and Society*, 3, 27–43.

Turri, E. (1982). *Dentro il paesaggio*. Verona: Bertani.

Viesti, G. (2001). *Come nascono i distretti industriali*. Roma-Bari: Laterza.

Zambon, I., Colantoni, A., Carlucci, M., Morrow, N., Sateriano, A., Salvati, L. (2017). Land quality, sustainable development and environmental degradation in agricultural districts: A computational approach based on entropy indexes. *Environmental Impact Assessment Review*, 64, 37–46.

Zambon, I., Colantoni, A., Cecchini, M., Mosconi, E. (2018). Rethinking sustainability within the viticulture realities integrating economy, landscape and energy. *Sustainability*, 10(2), 320.

Zambon, I., Monarca, D., Cecchini, M., Bedini, R., Longo, L., Romagnoli, M., Marucci, A. (2016). Alternative energy and the development of local rural contexts: An approach to improve the degree of smart cities in the Central-Southern Italy. *Contemporary Engeneering Science*, 9, 1371–1386.

Zangrandi, A. (2019). *Aziende pubbliche*, Milano: Egea.

Index

About the Authors

Sabato Vinci, MBA, Ph.D. is expert in Public Management and Governance. He is Research Fellow in Business Economics at University of Rome Tre, and Principal Researcher with coordination duties in the Public Governance and Health Management research line at Research Laboratory in Economics, Governance and Ethics of Organisations (LEGEA) on Department of Political Sciences, dealing mainly with State-owned enterprises, healthcare organisations, local public utilities, urban management and territorial development strategies. He held a Master in Business Administration (MBA) from LUISS Business School and a Ph.D. in the scientific field of Business Economics from University of Roma Tre. He collaborated in some research activities for "Fabio Gobbo" Industrial and Financial Research Group (GRIF) at LUISS Guido Carli University and for Council for Agricultural Research and Economics (CREA). He is former adjunct Professor at Master in Business Administration (MBA) and in other post-graduate Masters at Link Campus University. He is former adjunct Professor of Public Finance and Finance Management at National School of Administration (SNA) of the Presidency of the Council of Ministers of Italy. He carries out scientific and professional advisory activities for members of the European Parliament and for other Italian public institutions. He is a member of research groups and study commissions about management and economics matters. He has published many papers for national and international journals.

Luca Salvati, M.S., Ph.D. is expert in Regional Statistics, Demography, Economic Geography and Urban Planning. He is aggregate professor of Economic Statistics at University of Macerata – Department of Economics and Law, dealing mainly with institutional and economic statistics, statistics for the territory, theory of indicators, sustainable development and urban growth. He held a Ph.D. in Economic Geography and was a former staff researcher of the National Statistics Institute (ISTAT) and of the Council for Agricultural Research and Economics (CREA) in Italy. He has published more than 400 articles in English, 30 books, and many thematic essays, also with specific contributions on multivariate statistics and spatial analysis.

For Product Safety Concerns and Information please contact our EU
representative GPSR@taylorandfrancis.com
Taylor & Francis Verlag GmbH, Kaufingerstraße 24, 80331 München, Germany

www.ingramcontent.com/pod-product-compliance
Ingram Content Group UK Ltd.
Pitfield, Milton Keynes, MK11 3LW, UK
UKHW020156260926
475702UK00047B/308
* 9 7 8 8 7 7 0 2 2 1 6 6 5 *